# Robot Companions

## *MentorBots and Beyond*

**E. Oliver Severin**

**McGraw-Hill**

New York   Chicago   San Francisco   Lisbon   London   Madrid
Mexico City   Milan   New Delhi   San Juan   Seoul
Singapore   Sydney   Toronto

*The McGraw·Hill Companies*

**Library of Congress Cataloging-in-Publication Data**

Severin, E. Oliver
    Robot companions : MentorBots and beyond / E. Oliver Severin.
        p. cm.
    Includes index.
    ISBN 0-07-142212-9 (alk. paper)
    1. Robotics.    I. Title.

    TJ211.S439  2003
    303.48'34—dc22

                                        2003061580

1 2 3 4 5 6 7 8 9 0    DOC/DOC    0 9 8 7 6 5 4 3

ISBN 0-07-142212-9

*The sponsoring editor for this book was Judy Bass and the production supervisor was Sherri Souffrance. It was set in Sabon by Patricia Wallenburg. The art director for the cover was Anthony Landi.*

*Printed and bound by RR Donnelley.*

 This book was printed on recycled, acid-free paper containing a minimum of 50% recycled, de-inked fiber.

McGraw-Hill books are available at special quantity discounts to use as premiums and sales promotions, or for use in corporate training programs. For more information, please write to the Director of Special Sales, Professional Publishing, McGraw-Hill, Two Penn Plaza, New York, NY 10121-2298. Or contact your local bookstore.

# Contents

iii

ix

# Introduction

*Robot Companions* explores the growing need for companions for neglected children and lonely seniors. Starting with the basic need to share life with a being, and using the connection between humans and their technological brethren, this book explains how this relationship is being redefined as humans develop increasingly complex machines. The impetus to build machines that exhibit lifelike behaviors goes back centuries, but the technology to make it possible only recently emerged from the development labs. The goal is not simply to build machines that are like humans, but to alter our perception of the potential capabilities of robots. Our current attitude toward intelligent robots is simply a reflection of our own view of ourselves.

MentorBots are curious, simple, ubiquitous machines, with more in common with humans than one might think. The focus of much robot research has been to develop superhuman devices that operate at the highest intellectual levels. It might be better to make specialized robots that can perform only a handful of tasks, but do them well. The way toward a future in which robotic creatures work in tandem with, and even begin to resemble, humans is not a science fiction utopia, but a future in which people will need, and have, robotic companions with which to work.

*Robot Companions* challenges the view that human nature can be seen to possess the essential characteristics of a machine. Our instinctive rejection of that idea is itself a conditioned response. Once a direction is set, an exploration of present capabilities will be followed by a glimpse into future capabilities. For those who are so inclined, the fifth chapter of the book presents plans for building your own MentorBot and information on where to get the parts. Provocative, persuasive, and compelling, *Robot Companions* presents a vision of the future and our future selves.

■ **Figure I.1**  *One celebrity with robotic experience who also appeared with the Flo MentorBot was* Star Trek *Captain Kirk, William Shatner.*

■ **Figure I.2**  *Joseph Engelberger, the father of robotics and Unimation Founder, is pictured with Pearl, a robot developed at the University of Pittsburgh to attend and nurse the elderly. After the sale of Unimation, Engelberger spent a great deal of time with mobile robots. He was the Chairman of HelpMate Robotics Inc. until it was sold to Pyxis Corporation. Pearl is the second prototype of a personal robotic assistant developed by researchers from the University of Pittsburgh and Carnegie Mellon University. Also known as the "Nursebot," this personal robotic assistant could help elderly individuals who live on their own and suffer from mild cognitive disabilities or common physical ailments.*

# MentorBots Defined

IN 1921, THE TERM "ROBOT" WAS FIRST USED IN A PLAY called "Rossum's Universal Robots" by the Czech writer Karel Capek. In 1941, science fiction writer Isaac Asimov first used the word "robotics" to describe the technology of robots and predicted the rise of a powerful robot industry. In 1948, an influential artificial intelligence research paper, "Cybernetics," was published by Norbert Wiener.

In 1956, George Devol and Joseph Engelberger formed the world's first robot company. In 1961, the first industrial robot went online in a General Motors automobile factory in New Jersey; it was called UNIMATE. In 1963, the first artificial robotic arm to be controlled by a computer was designed. In 1969, the Stanford Arm became the first electrically powered, computer-controlled robot arm. In 1979, the Standford Cart crossed a chair-filled room without human assistance.

In science fiction many examples of intelligent, autonomous, and automated machines or robots exist. While the fantastic, sometimes frightening robots in books and movies are technologically complex, their interaction with humans is often depicted as unnatural. These fictitious robot characters are primarily designed to be entertaining.

Today's robots are neither mechanically nor structurally anywhere close to handling the same tasks as their fictitious counterparts. But we have graduated from remote-controlled contrivances to true autonomous mobile robots that can navigate through a maze on their own. There are even bipedal robots that can climb and descend stairs on their own.

The most desirable robots, from the general public standpoint, are just now coming out of the laboratories. These can do more than merely follow human commands, and they come in different shapes and sizes. These robots—MentorBots—can carry on a conversation, recognize their environment, identify the person they are talking to and make intelligent sense. And they can do a variety of useful tasks.

## MentorBot Types

The fictitious robot characters in movies are primarily designed to be entertaining, but these fictional robots have engendered a public fascination with mechanical entities. If we do not strive for the full functionality of such colorful creations as 3CPO and R2-D2 from the Star Wars movies, we will face a disappointing failure.

A number of applications for various MentorBots have already appeared. These applications are defined and explained in the next sections.

## Children's Playmate

Rather than recognizing cues from the child, the children's playmate MentorBots respond to the child with an expression of an emotion. Some are scientifically based, designed to realistically model the emotional states of a real animal, whereas others are more closely aligned with toys. These MentorBots are twitchy, cuddly creatures, designed to entertain a child. A robot that emulates the behavior of a real animal is not a new concept. These MentorBots are not, however, an exercise in artificial intelligence: Their motivations are not self guided, nor do they arise on their own without provocation. These MentorBots do not get hungry or thirsty; they are designed to facilitate research into the emotional education of children, rather than to explore an experimental implementation of a theoretical emotional model.

These are not as much commercial specialty toys as they are robots designed to model a realistic emotion system. Figure 1.1 shows Kasey the KinderBot, a Preschool MentorBot covered in Chapter 4. This MentorBot is an instructional robot that responds directly to the child without ulterior motives such as hunger or sleep to confuse his behavioral expression. When a child makes the MentorBot unhappy, it is because of something that the child has done, and not because the MentorBot has other unfulfilled needs.

Figure 1.2 shows six children's playmate MentorBots; the first four are covered in Chapter 4; the last two in Chapter 8.

■ **Figure 1.1**  *Kasey with child.*

In general, these affective toys and robots have begun to produce emotionally conscientious computer interfaces. The children's playmate MentorBot is in the unique position of being an educational computer toy. He not only recognizes and synthesizes emotional information, but also teaches the child about actions and consequences.

The MentorBot's reaction is indirectly produced by the child's emotional behavior. Thus, the connection can be made that if the child is hurting him, the MentorBot becomes unhappy, and vice versa. Specifically, through normal play experience, the child learns what she can do to make the MentorBot sad. Once the child makes the connection between her behavior and the MentorBot's response, she has learned and demonstrated what is referred to as emotional recognition. If the child can take the next developmental step and discover what she can do to make the MentorBot happy again, she has learned and demonstrated compassion and empathy. This is the beginning of emotional intelligence.

## Teacher's Assistant

In the near future, an astronaut in space will get a little assistant in the form of a robot that will follow the astronaut around, keeping track of

■ **Figure 1.2** *Playmate MentorBots. Top row, left to right, is Kasey the KinderBot, preschool playmate; AIBO ERS-31L, robotic dog; AIBO ERS-210A, companion dog. Bottom row, left to right, is NeCoRo, pet cat; PePaRo, personal robot; SDR-4X, bipedal entertainment robot.*

the astronaut's schedule. It will check computer files, monitor experiments, and update inventories. This assistant will also keep a log of conditions on the space shuttle and space station. It will check levels of oxygen and hydrogen and, if emissions reach critical levels, the assistant will let everyone know. This assistant is not humanoid; it's a robot that looks like a red grapefruit. It's packed full of sensors, miniaturized video equipment, wireless network equipment, and technology that allows the robot to understand spoken commands and reply with the same.

The teacher's assistant MentorBot should help the teacher perform his duties in the same way that the astronaut's assistant robot helps her to motivate and teach students good habits and to help grade papers.

Figure 1.3 shows RB5X, a teacher's assistant MentorBot covered in Chapter 4 along with other similar MentorBots.

What skills does the MentorBot teach? The age-specific study materials and optional school assembly presentations cover six key topics:

■  Self awareness
■  Interpersonal skills

■ **Figure 1.3** *RB5X teacher's assistant.*

- Decision-making
- Drug awareness
- Refusal skills
- Earth skills

A MentorBot teacher's assistant is effective for all students in Grades K–6 (ages 5–12). The lesson plans are designed to be easy for teachers and parents to use with all children in this age group.

## Teacher Substitute

The Teacher Substitute MentorBot is an educational tool capable of increasing a student's attention, comprehension, and motivation for learning in mathematics, science, language arts, social studies, and computer technology, at all grade levels. In a six-month controlled study at Washington Elementary, Las Cruces, New Mexico, sixth grade boys were found to increase their math comprehension by 40 percent, while girls of the same age and in the same period increased their math comprehension by 80 percent.

MDM is an interactive robot teacher, covered in Chapter 4. This MentorBot excels in helping students learn English as a second language (ESL). In fact, the robot can speak any language. Using universal phonemes, children increase their language skills by teaching the Teacher Substitute MentorBot to speak. Because the MentorBot does not usually speak exactly the way the student expects the first time, the student teaches the robot to pronounce the words correctly.

This is the principle by which the MentorBot has proven to be a powerful teaching tool. When teaching the MentorBot math, language arts, or social studies, students not only demonstrate their knowledge and identify their learning gaps, they develop self-esteem and problem solving skills, critical for success in today's world. Through this transformation, students take responsibility for their educational experiences and gain life-long motivation for learning. The MentorBot interacts with the class in ways that a computer simply cannot.

An article titled "Push My Button" describes how the MentorBot helped children increase their math and language skills, personal motivation, and interest in learning. The MentorBot can includes specific health and assertiveness skills that inspire children to protect themselves from cigarettes, alcohol, drugs, and other threats to their well-being. It can be used for multiple classes combined in an assembly and it is most effective for K-6 elementary schools.

A possible curriculum for the MentorBot includes:

- Motivating children to achieve their personal best by teaching responsibility, decision-making, and positive social skills.
- Teaching health and assertiveness skills to inspire children to protect themselves from cigarettes, alcohol, drugs, violence, and other threats.

The MentorBot's effectiveness is validated by five independent scientific studies, and more than 2,500,000 U.S. children have begun benefiting from MentorBot since its introduction in 1986. The MentorBot is recommended for children in Grades K–6 (ages 5–12).

After the MentorBot has completed its tasks, a human teacher should evaluate its impact on the students. Which types of instructions were the clearest? How could other directions be changed to improve the robot's understanding? What were some limitations imposed by only being able to verbalize instructions?

## Adult's Companion

The next generation of robot research will study the formation of relations between humans and humanoids. The exterior design of the robot will be integral to clarifying its diverse mechanical functions and asserting its autonomy as distinct from that of a mere object. This distinction will enable humans to interact meaningfully with the MentorBot and define a relationship between human and robot. Traditionally, robotic systems have necessitated, by the nature of their inner mechanisms, the dynamic by which the robot is appraised from the human standpoint. The exterior design of the MentorBot has transformed this notion by subverting the viewpoint so that the object is looking back at the human, thus creating a spatial dimension previously unexplored in the research of humanoid robot design.

Primarily, engineering factors have determined that the humanoid form is the most expedient in MentorBot design. For MentorBots to be successfully integrated into human society, their primary mobile functions will have to negotiate the same obstacles encountered by humans in daily life. In their role of aiding disaster relief and other potentially lethal employment, they will require the optimal mobility afforded by the human form in a human-designed landscape. Figure 1.4 shows AMI, a humanlike robot covered in Chapter 8.

■ **Figure 1.4**  *AMI adult MentorBot.*

However, when taking robotics a step further—defining the MentorBot as a coveted consumer item reflecting the desires and aspirations of its user—it is necessary to develop a system of aesthetic standards that will enable the user to form a psychological affiliation with the object.

Thus, the design must conform to the variable expectations of the user. In some cases the user may identify with the object on the basis of an aesthete: Is this form compatible with already existing notions of beauty? In other cases, the user may project upon the object his longing for something entirely different, whether it be basic companionship or the possession of yet another desirable gadget.

The relationship between MentorBots and humans is a factor that designers must explore deeply before robots can be successfully integrated into society. For the MentorBot to evolve from object to entity we need to address its genesis in purely human terms as we did with PINO, the humanoid robot who shares an ancestral link with Pinocchio, the puppet who aspired to be a boy.

## Executive's Gofer

In some ways, it is the ultimate personal assistant: It works while you sleep, shows up on time, and never complains about the hours. But it can't make independent decisions or a pot of coffee, and it needs specific instructions about what to do from its executive "Boss." Sound like a poor excuse for a robot? Meet today's Executive Gofer MentorBot—a glorified piece of hardware and software code to which you can delegate tasks you don't want to do yourself, like going for the mail or taking a document to another executive for signatures, or gathering all that is new and relevant to you and your job. This agent is no R2-D2, but the technology is slowly catching on as a way to speed up the "World Wide Wait."

As with many leading-edge technologies, few agree on just what constitutes a true intelligent agent. The consensus is that it acts on your behalf, whether as part of a computer network or as an autonomous machine. In today's first generation MentorBots, one goal is to have the robot move autonomously through an office building, reliably performing office tasks such as picking up and delivering mail and computer printouts, returning and picking up library books, and carrying recycling cans to the appropriate containers. Figure 1.5 shows ASIMO, an advanced humanoid covered in Chapter 6 along with other gofer humanoids.

The ultimate goal for the office MentorBot is a high-level reasoning process that will allow the MentorBot to efficiently handle multiple interacting goals and learn from its experience—a complete planning, executing, and learning autonomous robotic agent. Since the navigation module may get confused and report a success even in a failure situation, the MentorBot always verifies the location with a secondary test (vision or human interaction). If the MentorBot detects that it is not at the correct location, it updates its navigator domain knowledge to reflect the actual position, rather than the expected position.

In a similar manner, the MentorBot is able to detect when an action is no longer necessary. If an action unexpectedly achieves some other nec-

■ **Figure 1.5**  *ASIMO gofer.*

essary part of the plan, then that precondition is added to the state and the MentorBot will not need to subgoal to achieve it. Also, when an action accidentally cancels the effect of a previous action (and the change is detectable), the MentorBot deletes the relevant precondition and is forced to repeat it.

Take, for example, a coffee delivery scenario. The MentorBot picks up the coffee, adding the literal (has-item coffee) to its knowledge base and deleting the goal (pickup-item coffee room A). If the MentorBot is now interrupted with a more important task, it suspends the coffee delivery and does the other task. While doing the new task, the coffee gets cold, making the literal (has-item coffee) untrue. (The state change is detected by a manually encoded daemon.) When the MentorBot returns to the original task, it examines the next foreseeable action: (deliver-item coffee room B), discovers that a precondition is missing (it doesn't have the coffee), and repeats the first sequence.

In this manner, the MentorBot is able to detect simple execution failures and compensate for them. The interleaving of planning and execution reduces the need for replanning during the execution phase and increases the likelihood of overall plan success. It allows the system to adapt to a changing environment where failures can occur.

Figure 1.6 shows three executive gofer MentorBots. On the left is Robovie, an interactive robot; in the center is AMI, a humanlike robot; and on the right is H7, an action-integrated humanoid. All three are covered in Chapter 4.

■ **Figure 1.6**  *Gofer MentorBots.*

Observing the real world allows the MentorBot to adapt to its environment and to make intelligent and relevant planning decisions. Observation allows the planner to update and correct its domain model when it notice changes in the environment. For example, it can notice limited resources (e.g., battery); notice external events (e.g., doors opening/closing); or prune alternative outcomes of an operator. In these ways, observation can create opportunities for the planner and it can also reduce the planning effort. Real-world observation creates a more robust planner that is sensitive to its environment.

## Senior's Advisor

We have succeeded in helping people to live longer, but chronic illnesses and disabilities often prematurely force elderly persons into personal care homes. The Senior's Nurse MentorBot has the potential to help older persons remain independent longer and live a better quality of life.

MentorBots are being developed for taking care of the elderly. One called Wakamaru is built on a small mobile platform with navigational software and infrared and sonar sensors that allow it to move safely while tracking an elderly companion's movements. Users operate Wakamaru through gestures recognized by machine vision, or by giving the robot vocal commands. A top priority is to make Wakamaru more responsive to the abilities of the average person. The simpler the robot; the more effective a tool it will become.

The purpose of a Senior Advisor MentorBot (Figure 1.7) is to provide a mobile, personal-service robot to assist elderly people suffering from chronic disorders. Currently under shakedown, Pearl is an autonomous

■ **Figure 1.7** *Advisor Pearl.*

mobile robot that "lives" in the private home of a chronically ill elderly person.

This robot provides a research platform to test out a range of ideas for assisting elderly people, such as:

- **Intelligent reminding.** Many elderly patients have to give up independent living because they forget to visit the restroom, take medicine, drink, or see the doctor. This project explores the effectiveness of a robotic reminder that follows people around (and cannot get lost).
- **Telepresence.** Professional caregivers can use the robot to establish a "telepresence" and interact directly with remote patients, thus making many doctor visits superfluous. The robot connects patients with caregivers through the Next Generation Internet (NGI).
- **Data collection and surveillance.** A range of emergency conditions can be avoided with systematic data collection (e.g., certain types of heart failures). This reason alone can make service robots succeed in the home care business.

- **Mobile manipulation.** Arthritis is the main reason that elderly persons give up independent living. A semi-intelligent mobile manipulator that integrates robotic strength with human senses and intellect can overcome barriers in manipulating objects (refrigerator, laundry, microwave) that currently force patients to move into assisted living facilities.
- **Social interaction.** A huge number of elderly people are forced to live alone, deprived of social contacts. The MentorBot project seeks to explore whether robots can take over certain social functions.

If successful, this project could change the way we deliver healthcare to the ever-growing contingent of elderly people, and it could significantly advance the state of the art in mobile service robotics and human robot interaction (Figure 1.8).

You can have a conversation with Pearl, but only on the limited topics chosen by our research team. Ideally, Senior's Nurse MentorBot should be able to discuss any subject pertinent to its job. These robots could make their way into homes within the next decade: With the elderly population rapidly increasing, personal robotic assistants may become as popular a fixture as walkers and canes in the homes of elderly persons.

- **Figure 1.8** *Upper left, AIBO, robotic dog; upper right, NeCoRo, pet cat, both covered in Chapter 4. Lower left, PePaRo, advanced personal robot; lower right, SDR-4X, small advanced biped, both covered in Chapter 8.*

Perhaps it is time for robotics to focus on being a part of a solution to one of the major crises of our time—caring for our aging population. The technology required to address the needs of the elderly has been developed and is primed to continue to grow if we choose to invest in it. Although the need for a realistic plan to care for the increasing number of elderly is not new, it is becoming more critical as our population continues to age.

## Beyond MentorBots

The robot systems described in the following sections take the initial idea of the MentorBot a step farther, into the realm of humanoid robotics.

### Children's Nanny

The kids love her—she'll play with them anytime and do whatever they like. She's part of the family. She's always there for the children, not like the au pairs of the last century who were around for only six weeks at a time. She does everything from changing diapers with her soft, dexterous fingers to taking grandmother grocery shopping and keeping her company, listening to her same stories over and over, without ever being bored about the time grandma slept with Elvis Presley in the '50s.

Figure 1.9 shows Wakamura, a caregiver robot covered in Chapter 4. Here, the temporal synchronization of two agents—a robot and a human—can be exploited for teaching purposes, so that the robot's actions are influenced by the way the human responds to its actions.

Getting the interaction dynamics right between infant and nanny seems to be a central step in the development of social skills, since they require coordinating one's external and internal states with another agent. A simple analysis of this interaction between robot and child is:

- The robot understands the conditions of the child based on interactive sequences, actions, voice, and timing reactions.
- The child understands the robot depending on the human social context.

This collaboration indicates a very interesting idea that the value and significance of robots can be discovered in communication with users. Therefore, it is necessary to design the relationship between the robot and user, as well as the robot itself.

■ **Figure 1.9** *Nanny Wakamaru.*

A Nanny MentorBot must be designed so that people feel comfortable with it and do not need to learn any new skills to have gratifying interactions with it. For this type of social experience to happen, the robot must respect the basic rules of human interaction, such as respecting another's interpersonal space, and project the basic gestural cues used in human interactions, such as those which designate where the robot's attention is focused.

Paradoxes arise as soon as you permit your Children's Nanny MentorBot to use ordinary commonsense reasoning. Here are some of the effects.

1. Keeping house becomes trivial. The house or apartment will be clean and neat and things will be kept put away. We suppose that shopping also becomes trivial, either by sending the robot to the store or, more likely, by an automatic delivery system. The latter is an easier technology and will probably be available sooner than robots.

2. How much robots will affect life depends critically on the extent to which they can be used to take care of babies and children. To be acceptable in homes with children, they must at least be smart enough not to step on them or otherwise injure them even if their

duties don't include child care itself. The next step is watching over sleeping babies.

3. The robot must be able to detect emergencies, remove a child from fire or other danger, and call for medical help. This should be a small step for a robot that can physically handle a baby or child without danger. Since the robot can remain alert to an extent that humans cannot, the level of physical safety will promptly become higher than is presently achieved even with parents present.

4. The social effects of household robots will be profound, but the situation isn't entirely unprecedented: We can compare it with that of Victorian upper- and middle-class households that had servants. However, robots will be more universally available.

## Adult's Servant

An Adult Servant MentorBot indulges in some wishful thinking because the artificial intelligence problems involved in making a general-purpose robot servant include unsolved conceptual problems that require basic research. It is not possible to say when these problems will be solved, although the mechanical engineering problems seem within "development range" (i.e., solutions could be financed once AI advances make it worthwhile). Figure 1.10 shows PINO, a humanoid covered in Chapter 4.

■ **Figure 1.10**  *Servant PINO.*

Let us suppose that general-purpose household robots become available ten years from now. How would they be used, and how would they affect society? We are assuming they would be affordable and universally available, because if they are truly general purpose, then they can also be used to make more of them.

We imagine that household robots will have the capabilities of people but won't be people. (We discuss later the psychological characteristics that will prevent robots from being appropriate objects of either sympathy or blame.)

A robot can work 24 hours a day, but at first, people will use them only to take over activities that they have previously done themselves. A higher standard of cleanliness will become the norm: No dust will be allowed to accumulate anywhere, minor dents will be fixed immediately, windows will be cleaned daily.

We can also imagine MentorBot porters. People will be inclined to take a robot along when they go somewhere to carry packages, clothing, and or even a favorite chair and a book or TV in case of boredom. Finally, the owner may want the robot to carry him in case he gets tired. At that level, one robot won't be enough.

When a car is used, whether driven by a human, robot, or built-in computer, robots going along for the ride may stand on an outside rack like footmen on a coach. (Intellectuals will surely express disgust with the unnecessary luxury involved. Perhaps some will propose *sumptuary laws* [as the ancient Romans called them] to curb the excesses: No-one will be allowed to bring more than six robots into an opera house, even if they are programmed to stack themselves in the cloak room.)

Women often want to look their very best at any social occasion, and robots will extend this capability. One porter robot will be a closet. Just before making an entrance a woman may step into the closet, which will take off her clothes, clean her to the utmost, put on clothes most suited to the occasion, and fix her hair in the appropriate style. Indeed, after going into the house, theater, or restaurant, the robot may look at the others present and suggest changes in appearance. She'll duck out and the closet will make the appropriate changes in dress and make-up.

There may be glitches. For example, a household robot may not want its masters to get anything for themselves or to put anything away. "Let me do it; then I'll remember where it is when you ask me for it again." The robot will be especially assertive with children, who very often forget where they put things.

Programming robots with human motivations and desires should be avoided so that their behavior will not tempt people, especially children, to either like or dislike them. Otherwise, people will be tempted to give them a status in society, and this will lead to all sorts of science fiction stories being acted out. What consciousness a robot will need is discussed technically in *Making Robots Conscious of Their Mental States* (Stanford University paper by John McCarthy) Perhaps robots should take themselves apart when not in use and reassemble into different configurations when summoned for use.

Finally, what about unemployment? Won't robots take jobs away from people? Many past inventions have increased productivity and thereby reduced employment in particular industries. For example, the fraction of the American population who are farmers today is about 2 percent; once it was 90 percent. The number of underground coal miners is down by 90 percent since the end of World War II.

However, there aren't a fixed number of jobs, and the overall rate of unemployment has been unconnected with increased productivity, both within a country and between countries. Indeed, the largest rates of unemployment are in backward countries. What if eventually robots could do all work?

## Executive's Advisor and Other MentorBots

Researchers at Carnegie Mellon University are working on an Executive's Advisor MentorBot named Doc Beardsley, a descendant of the mechanical humans and beasts that rang bells and performed other actions as parts of the clocks of medieval European cathedrals.

Modern science, however, has carried Doc far beyond these ancient automata, endowing him with the ability to see, understand spoken words, and carry on a conversation. Several layers of software drive Doc's apparent wit. Synthetic interview software, which includes speech recognition abilities, allows Doc to react to spoken questions. The technology, developed at Carnegie Mellon for use with video characters, gives a character sets of lines to deliver on given topics. This allows Doc to give appropriate answers to questions that match an anticipated query closely enough.

If the question hasn't been anticipated, another layer of software takes over. A discussion engine tracks the questions and answers during a conversation and allows Doc to make relevant comments by keying off individual words even if he doesn't understand a specific question. And if that doesn't work, the discussion engine tosses the conversation back to the questioner. Figure 1.11 shows AMI, a humanlike robot, covered in Chapter 4.

The discussion engine first tries to deliver a comment that is still relevant based on the individual keywords that can be found in the text. Failing that, the robot gives a random comment that either pretends to reflect what is being discussed to try to keep the conversation going, or transfers the onus of the conversation back to the guest.

From the entertainment perspective, the ultimate goal is creating synthetic robotic characters that seem to possess dramatic human qualities, like a sense of humor, comic timing, personal motivations, and improvisation. When an audience can get so engrossed in interacting with the MentorBot's dialogue that the technology is completely forgotten, we can progress to the next step: improving the character by adding skin and a costume, building a set and props, creating a show, building puppeteering controls for the props, and writing software for producing other shows.

■ **Figure 1.11**   *Advisor AMI.*

## Summary

A robot has traditionally been viewed as a tool: a device capable of performing tasks on (human) command. Although this "robot as tool" approach suffices for some domains, it is suboptimal for tasks that require significant human–robot collaboration or interaction. Such tasks include elderly care giving, assisting the physically disabled, tour guiding, and collaborative exploration.

To support these applications, recent significant effort has been made to develop robots that function more "naturally," and can interact on human terms. Instead of serving as mere tools, these "social robots" are designed to operate more like partners, if not peers. These robots exhibit the adaptability and flexibility to drive interactions with a wide range of humans, from novices to experts, children to adults. They are capable of social interaction and may express emotions, may communicate with natural language or high-level dialogue, and may exhibit distinctive personality. We have selected MentorBots to identify this class of autonomous robot.

The goal here is to make robots more useful and acceptable by enabling them to interact with humans using social rules and conventions. This includes such rules as how to pass people in hallways in a socially acceptable manner, ride in elevators, and enter and wait in line. In conjunction with members of the drama department, we are starting a project to give a robot a personality and have it converse with people. The goal is to develop a robot receptionist that is both useful and entertaining.

Inspired by infant social development, psychology, ethnology, and evolution, this work integrates theories and concepts from these diverse viewpoints to enable MentorBots to enter into a natural and intuitive social interaction with human caregivers and to learn from them, reminiscent of parent–infant exchanges. To do this, MentorBots perceive a variety of natural social cues from visual and auditory channels and deliver social signals to the human caregiver through gaze direction, facial expression, body posture, and vocal babbles.

MentorBots have been designed to support several social cues and skills that could ultimately play an important role in socially situated learning with a human instructor. These capabilities are evaluated with respect to the ability of naive subjects to read and interpret the robot's social cues, the robot's ability to perceive and appropriately respond to human social cues, the human's willingness to provide scaffolding to facilitate the robot's learning, and how this produces a rich, flexible, dynamic interaction that is physical, affective, social, and affords a rich opportunity for learning.

**2**

# MentorBot Characteristics

SINCE THE DAWN OF THE COMPUTER AGE, OCEANS OF INK have been spilled writing about intention and conscious states and how to define them, and what sort of organisms or machines might have certain of these qualities. The battles continue to rage about whether a machine could ever approach consciousness in the way that we understand it and make meaning the way we do. Oddly enough, in the search for the truth of the matter, both camps have overlooked an obvious strategy: interviewing a computer and asking its opinion.

Robots are an intriguing technology because of their potential to play a rich and rewarding part in our physical and social world. Taking inspiration and guidance from the science of animal and human behavior, a MentorBot should be a capable and appealing robot creature that can engage us, communicate with us, and give to and take from us on our terms. This is not only an engineering endeavor, but a project to gain scientific insight into the mechanisms that underlie human and animal capabilities, and to develop a science of human–robot interaction. Given the multidisciplinary nature of this endeavor, a MentorBot should satisfy a wide variety of characteristics including:

- Novel mechanical designs
- Ability to quickly determine position and orientation
- Ability to find home base and recharge without human assistance
- Ability to traverse varied surfaces, such as rug–tile transition
- New sensing and actuator technologies
- Active multimodal perceptual systems

- Speech recognition and synthesis
- Expressive motion and skillful motor control
- Social learning
- Psychological modeling
- Human–robot interaction

Once deep philosophical conundrums were important only to people who had nothing better to do than wonder why they wondered; however, in recent years, these problems have developed into a host of potential practical difficulties in the progress of computer science. Essentially, these questions boil down to this:

- Can machines think?
- Do people with mortgages have free will?
- What is the meaning of life?

Many people insist that no machine can really think. "Yes," they say, "machines can do many clever things. But all of that is based on tricks, just programs written by people to make those machines obey preconceived rules. The results are useful enough—but nowhere in those cold machines is there any feeling, meaning, or consciousness. Those computers simply have no sense that anything is happening."

## Consciousness

Whatever consciousness might be, it has a quality that categorically places it outside the realm of science: namely, a subjective character that makes it utterly private and unobservable. Why do so many people feel that consciousness cannot be explained in terms of anything science can presently do?

Instead of arguing about that issue, let's try to understand the source of that skeptical attitude. Many people maintain that, even if a machine were programmed to behave in a manner indistinguishable from that of a person, it still could not have any subjective experience. Now isn't that a strange belief? Unless you were a machine yourself, how could you possibly know such a thing? As for "subjectivity," consider that *talking* about consciousness is a common, objective form of behavior. Therefore, any machine that suitably simulated a human brain would have to produce that behavior. Then, wouldn't it be curious for our artificial entity to falsely claim to have consciousness? For if it had no such experience, then how could it possibly know what to say? A classic question in philosophy is asking for proof that our friends have minds; perhaps they are merely unfeeling machines.

They used to say the same about automata vis-a-vis animals. "Yes, those robots are ingenious, but they lack the essential spark of life." Biology then, and psychology now: Each was seen to need some essence that wasn't mechanical.

## What Makes a Person Human?

People have even less to say about questions:

- How do you know how to move your arm?
- How do you choose which words to say?
- How do you recognize what you see?
- How do you locate your memories?
- Why does *seeing* feel different from *hearing*?
- Why does *red* look so different from *green*?
- Why are emotions so hard to describe?
- What does "meaning" mean?
- How does reasoning work?
- How do we make generalizations?
- How do we get (make) new ideas?
- How does commonsense reasoning work?
- Why do we like pleasure more than pain?
- What are pain and pleasure, anyway?

Researchers in artificial intelligence (AI) discovered a wide variety of ways to make machines do pattern recognition, learning, problem solving, theorem proving, game-playing, induction and generalization, and language manipulation (Figure 2.1). To be sure, no one of those programs seemed much like a mind, because each one was so specialized. But now we're beginning to understand that there may be no need to seek a single magical "unified theory" or any single and hitherto unknown "fundamental principle," because thinking may instead be the product of many different mechanisms, competing as much as cooperating, and generally unperceived and unsuspected in the ordinary course of our everyday thought. Therefore, there is no central principle, no basic secret of life. Instead, what we have are huge organizations, painfully evolved, that manage to do what must be done by hook or crook, by whatever has been found to work.

■ **Figure 2.1** *Humanoid and human.*

## Assume the Same for the Mind

Some machines are already potentially more conscious than people, and further enhancements would be relatively easy to make. However, this does not imply that those machines would thereby, automatically, become much more intelligent. It is one thing to have access to data, but another thing to know how to make good use of it. Knowing how your pancreas works does not make you better at digesting your food.

So consider now, to what extent are you aware? How much do you know about how you walk? It is interesting to tell someone about the basic form of biped locomotion: You move in such a way as to start falling, and then you extend your leg to stop that fall. Most people are surprised at this, few know which muscles are involved, and a few people, don't even know which muscles they possess. In short, we are not much aware of what our bodies do. We're even less aware of what goes on inside our brains.

The world of science is still filled with mysteries. We're still not sure of how the sun produces all its heat. We do not know precisely where our early ancestors evolved. We can't yet say to what extent observing violence leads to crime. But questions like those do not evoke assertions of futility. Instead, we try harder to detect more neutrinos, find more fossils, or perform more thorough surveys. However, in certain areas of thought, more people take a different stance about the nature of our ignorance. They proceed to work hard, not toward finding answers, but toward trying to show that there are none.

## Consciousness Needs for Robots

In some respects, it is easy to provide computer programs with more powerful introspective abilities than humans have. A computer program can inspect itself, and many programs do this in a rather trivial way. They compute check sums to verify that they have been read into computer memory without modification.

It is easy to make available for inspection by the program the manuals for the programming language used, the manual for the computer itself, and a copy of the compiler. A computer program can use this information to simulate what it would do if provided with given inputs. It can answer a question like: "Would I print "YES" in less than 1,000,000 steps for a certain input?" A finite version of Alan Turing's argument that the *halting problem* is unsolvable tells us that a computer cannot, in general, answer questions about what it would do in $n$steps in less than $n$steps. If it could, we (or a computer program) could construct a program that would answer a question about what it would do in $n$steps and then do the opposite (Figure 2.2).

Unfortunately, these easy forms of introspection are not especially useful for intelligent behavior in many commonsense information situations.

We humans have rather weak memories of the events in our lives, especially of intellectual events. The ability to remember its entire intellectual history is possible for a computer program and can be used by the program to modify its beliefs on the basis of new inferences or observations. This may prove very powerful.

To do the tasks we will give them, a robot will need at least the following forms of self-consciousness, or the ability to observe its own mental state. When we say that something is *observable*, we mean that a suitable *action* by the robot causes a sentence and possibly other data structures to give the result of the observation to appear in the robot's consciousness.

We will give tentative formulas for some of the results of these observations.

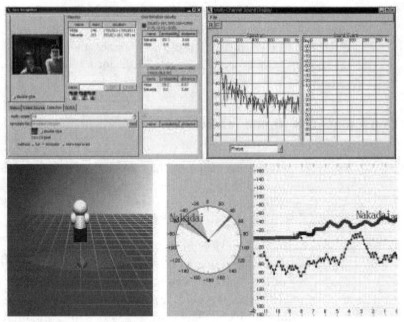

■ **Figure 2.2**  *Measuring consciousnesses.*

- Observing its physical body, recognizing the positions of its effectors, noticing the relation of its body to the environment and noticing the values of important internal variables, such as the state of its power supply and communication channels.

- Observing that it does or doesn't know the value of a certain term, such as the telephone number of a certain person. Observing that it does know the number or that it can get it by some procedure is likely to be straightforward.

- Deciding that it doesn't know and cannot infer the value of a telephone number should motivate the robot to look in the phone book or ask someone.

- Keeping a journal of physical and intellectual events so that it can refer to its past beliefs, observations, and actions.

- Observing its goal structure and forming sentences about it. Notice that merely having a stack of subgoals doesn't achieve this unless the stack is observable and not just obeyable.

- Having the *intention* to perform a certain action. It may later infer that certain possibilities are irrelevant in view of its intentions. This requires the ability to observe intentions.

- Observing how it arrived at its current beliefs. Most of the important beliefs of the system will have been obtained by no monotonic reasoning, and therefore are usually uncertain. It will need to maintain a critical view of these beliefs, (i.e., believe meta-sentences about these beliefs will aid in revising them when new information warrants doing so). It will presumably be useful to maintain a pedigree for each system belief so that it can be revised if its logical ancestors are revised. Reason maintenance systems maintain the pedigrees, but not in the form of sentences that can be used in reasoning. Neither do they have introspective subroutines that can observe the pedigrees and generate sentences about them. Not only pedigrees of beliefs, but other auxiliary information should either be represented as sentences or be observable in such a way as to give rise to sentences.

- Regarding its entire mental state up to the present as an object, the ability to *transcend* one's present context and think about it as an object is an important form of introspection, especially when we compare human and machine intelligence.

- Knowing what goals it can currently achieve and what its choices are for action. We claim that the ability to understand one's own choices constitutes *free will*.

## Understanding and Awareness

We do not offer definitions of understanding and awareness. Instead we discuss which abilities related to these phenomena robots will require. Consider fish swimming. Fish do not understand swimming, in the following senses:

- A fish cannot, while not swimming, review its previous swimming performance so as to swim better next time.
- A fish cannot take instruction from a more experienced fish in how to swim better.
- A fish cannot contemplate designing a fish better adapted to certain swimming conditions than it is.
- A human swimmer may understand more or less about swimming than a fish.

We contend that intelligent robots will need understanding of how they do things in order to improve their behavior in ways that fish cannot. Understanding is not an all-or-nothing quality.

Consider a robot that swims. Besides having a program for swimming with which it can interact, a logic-based robot needs to use sentences about swimming to give instructions to the program and to improve it.

The *understanding* a logical robot needs, then, requires it to use appropriate sentences about the matter being understood. The understanding involves both getting the sentences from observation and inference and using them appropriately to decide what to do.

*Awareness* is similar. It is a process whereby appropriate sentences about the world and its own mental situation come into the robot's consciousness, usually without intentional actions. Both understanding and awareness may be present to varying degrees in natural and artificial systems. The swimming robot may understand some facts about swimming and not others, and it may be aware of some aspects of its current swimming state and not others.

These are only some of the needed forms of self-consciousness. Research is needed to determine their properties and to find additional useful forms of self-consciousness.

## Human Brain Immolation

Similarly, we can ask the extent to which we're aware of the words we speak. At first one thinks, "Yes, I certainly can remember that I just pronounced 'the words we speak'." But to what extent are we aware of the process that produced those particular words? Why, barely at all! We have to employ linguists and lifetimes of research even to discover the simplest aspects of the language production process.

Finally, I can ask you a question like, "Can you tell me what you are thinking about?" You might list the names of some subjects or concerns that were recently in your mind, and sometimes you can describe a bit of the trains of thought that led to them. These kinds of answers clearly feed upon memories of recent brain activities. But every such answer seems incomplete, as though the act of probing into any one of those memories interferes with subsequently reaching any others. One cannot think of any aspect of consciousness that could operate without making use of short-term memories. This suggests that the term "consciousness" is used in connection with what the brain was being used for in its most recent state.

This raises the question of the extent to which such memories might really exist inside our brains. Clearly there is a problem: If the same neural network has been used recently for only a single purpose, then it may still contain substantial information about what it recently did. But if it was used for several things, then most of those traces will have been overwritten—unless some special hardware has been evolved for maintaining such records. For a modern computer, there is much less of a problem with this because we can write programs to store such records inside the machine's "general purpose memory."

Certainly, a certain degree of consciousness—in the sense of access to such records—is necessary for a person (or machine) to be intelligent. Also, after one has successfully solved a difficult problem, one wants to "assign credit" to those actions that actually helped. This may involve a good deal of analysis—in effect, thinking about what you've recently done—which clearly requires good records.

## Thinking

Ruling out the notion that machines can think stands upon a single and simple mistake. It overlooks the possibility of including systems "that are mistaken about mathematics to some degree, or systems that can change their minds." By inadvertently ruling such machines out, you've simply begged the question whether human mathematicians can be kinds of machines—because people do indeed change their minds, and can indeed be mistaken about some parts of mathematics. An entire generation of logical philosophers has thus wrongly tried to force their theories of mind to fit the rigid frames of formal logic. In doing so, they cut themselves off from the powerful new discoveries of computer science. Yes, it is true that we can describe the operation of a computer's hardware in terms of simple logical expressions. But no, we cannot use the same expressions to describe the meanings of that computer's output, because that would require us to formalize those descriptions inside the same logical system. And this, I claim, is something we cannot do without violating that assumption of consistency.

### Can Robots Outsmart Humans?

Perhaps the most important aspect of how we humans work is how we ask ourselves (not necessarily by using words) what problems we have seen before that most closely resemble the present case and how we managed to deal with them. We make mistakes and then sometimes manage to learn from them. We somehow employ capabilities for retrieving and then manipulating descriptions of some of our earlier mental activities. Now, notice how self-referent this is.

Often, when you work on a problem, you consider doing some certain thing—but before you actually carry that out, you often inquire about yourself, about whether you actually will be able to carry it through.

Solving problems isn't merely applying rules of inference to axioms. It involves making heuristic assessments about which aspects of the problem are essential and which of one's own abilities might be adequate to deal with them. Then, whatever happens next arouses various feelings and memories of situations that seem similar and methods that might be appropriate. Is this done by some kind of nonphysical magic, or it is accomplished by the huge and complex collection of knowledge-based representations and pattern-matching processes that we all regard as common sense?

## Autonomous Spatial Learning

To act in an unknown and continuously changing environment, an autonomous robot must be able to react instantaneously to changes and unexpected events to avoid collisions and to update its maps. Successful navigation requires that the robot react primarily on its immediate sensory information and secondarily on its internal mapping of the spatial layout of the environment.

An experimental mobile robot has been developed and constructed equipped with a number of complementary sensory systems. A video camera is mounted on a movable head that also contains a pair of microphones. Ultrasonic sensors are located around the body of the robot and a set of tactile sensors (whiskers) and a bumper are used to detect obstacles at a short range.

The project aims at developing the attention and navigation systems of the robot to include vision for spatial orientation. The choice of vision is natural, since this modality contains the richest information for this task. The problems we are studying include the automatic recognition of visual landmarks and reactions towards changes in the environment, as well as the production of linguistic output on unexpected events. The solutions to these problems are highly dependent on the behavior of the robot and not only on its perceptual abilities. In this view, the main problem of visual navigation is not vision itself, but rather the behavior that makes vision useful.

We have performed extensive computer simulations of reactive navigation and learning based on other modalities and developed algorithms for visual place recognition and motion detection. A simple form of visually based obstacle avoidance has already been implemented successfully on the robot, together with a tactile reactive control system. We have also studied the connection between visual input and linguistic output, and developed a neural network–based system that is able to learn spatial relations between objects and produce elementary linguistic output.

Our theoretical aim is to develop learning methods for autonomous robots that can construct control strategies based directly on their sensory and locomotion abilities. Instead of using a prespecified map, like a CAD design, our goal is to let the robot construct its own map from its sensory inputs in a form suitable for its own actions. Furthermore, the maps constructed should not depend on a specially prepared environment or artificial landmarks. We are using natural visual input from the video camera.

Because a large database exists on spatial orientation in biological systems, this research field is one of the most promising areas for cognitive technology inspired by biological systems. We have previously studied biologically inspired architectural principles for the construction of autonomous robots. This work included a study of goal-gradients as a general representational tool.

In a longer perspective, autonomous robots equipped with spatial learning have immense potential for industrial applications. A system developed along the lines of our project will be useful in new types of automatic industries (e.g., in auto carriers). Furthermore, such systems can be used in applications for the physically disabled, because autonomous mobile robots can function in a home environment that is not specially designed for robots. The project is highly interdisciplinary and combines cognitive, language, and neural network technology with autonomous systems.

## Responding Logically

Paradoxes arise as soon as you permit your machine to use ordinary commonsense reasoning. For example, troubles appear as soon as you try to speak about your own sentences, as in "this sentence is false" or "this statement has no proof" or in "this barber shaves all persons who don't shave themselves." The trouble is that when you permit "self reference," you can quickly produce absurdities. Now you might say, "Well then, why don't we redesign the system so that it cannot refer to itself?" The answer is that the logicians have never found a way to do this without either getting into worse problems or producing a system too constrained to be useful.

Isaac Asimov, the greatest robotics writer of all time, wrote his three laws in a short story called "Runaround," which was published by Street and Smith Publications, Inc. in 1942. The three laws were stated as follows:

- A robot may not injure a human being, or, through inaction, allow a human being to come to harm.
- A robot must obey the orders given it by human beings except where such orders would conflict with the First Law.
- A robot must protect its own existence as long as such protection does not conflict with the First or Second Law.

Building a sociable machine is also key to building a smarter machine. Most current robots are programmed to be very good at a specific task—say, navigating a room—but they can't do much more.

Dr. Cynthia Breazeal, while working at the MIT Artificial Intelligence Lab, is developing the Kismet system. "I'm building a robot that can leverage off the social structure that people already use to help each other learn. If we can build a robot that can tap into that system, then we might not have to program in every piece of its behavior."

### Inspired by Children

The work on Kismet, which began in 1997, is heavily inspired by child developmental psychology. "The robot starts off in a rather helpless and primitive condition, and requires the help of a sophisticated and benevolent caretaker to learn and develop," Dr. Breazeal said. Even Kismet's physical features—which include big blue eyes, lips, ears, and eyebrows—are patterned after features known to elicit a care-giving response from human adults (Figure 2.3).

■ **Figure 2.3** *Inspired by children.*

The eyes are actually sensors that allow the robot to glean information from its environment, such as whether something is being jiggled next to its face. Kismet can then respond to such stimuli—say move its head back if an object comes too close—and communicate a number of emotionlike processes (such as happiness, fear, and disgust). A human wears a microphone to talk to the robot, which also has microphones in its ears. The latter will eventually be used for sound localization.

The robot's features, behavior, and "emotions" work together so that it can "interact with humans in an intuitive, natural way," Breazeal said. For example, if an object is too close for the robot's cameras to see well, Kismet backs away. "This behavior, by itself, aids the cameras somewhat by increasing the distance between Kismet and the human," Breazeal said. "But the behavior can have a secondary and greater effect through social amplification. A withdrawal response is a strong social cue for the human to back away."

Kismet, she noted, is the exact opposite of Hal, the menacing robot in the movie *2001: A Space Odyssey*. "Hal is simply a glowing red light with no feedback as to what the machine is thinking. That's why it's so eerie. Kismet, on the other hand, both gives and takes feedback to communicate. I think people are often afraid that technology is making us less human. Kismet is a counterpoint to that—it really celebrates our humanity. This is a robot that thrives on social interactions."

### Advanced Learning

Another project goal for Kismet is to develop learning methods that make generalizations possible without giving up the control view of behavior. New places have many similarities with old situations, and the robot should generalize from previously encountered situations. The central problem here is to represent actions and situations on multiple levels. Such ability is in many respects similar to *chunking* in SOAR. But, as in most traditional systems using chunking, the operators in SOAR do not constitute control strategies and cannot be directly used to control the actions of a robot. This is something researchers hope to accomplish using selective learning methods in neural networks.

"Higher level" representations also aid the construction of a forward model of the environment. This gives the robot the ability to train and retrain its reinforcement learning system faster than without such a model. The robot constructs an inner world in which actions can be tried out before committing them to unforgiving reality. This is especially important when the environment has changed and a large number of updates are necessary to establish the new goal gradient.

There are two long-term goals of the project. The first is to develop models into a complete and self-contained system that can be used as the basis for a robot product.

The second long-range goal is theoretical and involves the extension of the model to more complex situations and learning tasks. In the design of a more general model, one must take into account not only that is the environment unknown, but also the nature of the problems that the robot encounters.

Consequently, the robot must create a problem space before it can search for a solution. This task would be impossible to handle if the robot were not able to reapply parts of it knowledge from other problem spaces. By being able to generalize from a familiar situation to an unknown one, the robot does not need to construct the problem space from scratch.

To accomplish such dynamic learning, one cannot describe the process of knowledge acquisition in a static metasystem. If this were possible, the acquisition would be restricted to the frames of that system. Any event that cannot be interpreted by the metasystem can simply not be taken into account. It follows that there will be a predetermined limitation on the knowledge that the robot can acquire.

**34**

### Making It Lifelike

To make Kismet as lifelike as possible, Breazeal and colleagues have not only incorporated findings from developmental psychology, but have also invited the comments of cartoon animators. "How do you make something that's not alive appear lifelike? That's what animators do so well," Breazeal explained.

The proverbial wizard behind the curtain (or in this case, wall) is a bank of fifteen computers, that process software programs that allow the robot to perceive its environment, analyze what it finds, and react.

In experiments over the last year or so, the researchers have been exploring how the robot interacts with people who aren't familiar with it. Are Kismet's actions and emotions understandable? Do people use those actions as feedback to adjust their own responses? Conversely, is the robot correctly "reading" its visitors?

Results to date are encouraging. For example, many people who've met Kismet have told Breazeal that the robot has a real presence. "It seems to really impact them on an emotional level, to the point where they tell me that when I turn Kismet off, it's really jarring. That's powerful. It means that I've really captured something in this robot that's special. That kind of reaction is also critical to the robot's design and purpose."

Once Kismet's social skills are optimized, "we can move on to other forms of learning," Breazeal said. In early work to that end, the researchers are teaching Kismet how to use its voice to negotiate the social world. "We want it to be able to get people to do things for it, much like a very young child."

The algorithms key to this will allow the robot to "learn" by trial and error. When it first attempts a task, it won't be very good. The robot will "remember" its mistakes, however, and make incremental improvements as it goes along. It can then apply what it's learned to completing the same task under different conditions.

# Communication

Sociable humanoid robots pose a dramatic and intriguing shift in the way one thinks about the control of autonomous robots. Traditionally, autonomous robots are designed to operate as independently and remotely as possible from humans, often performing tasks in hazardous and hostile environments. A new range of application domains (domestic, entertainment, health care), are driving the development of robots that can interact and cooperate with people and play a part in their daily lives. They can communicate in a manner that supports the natural communication modalities of humans. Examples include facial expression, body posture, gesture, gaze direction, and voice. The ability for people to communicate naturally with these machines is important. However, for suitably complex environments and tasks, the ability for people to teach these robots intuitively will also be important. Social aspects enter profoundly into both of these challenging problem domains.

## Hearing and Speaking

Humans hear a mixture of sounds, not a single sound source. *Automatic speech recognition* (ASR) assumes that the input is a voiced speech, and this assumption holds as long as a microphone is set close to the mouth of a speaker. The speech recognition community develops robust ASR to make sure this assumption holds over on wider fields.

*Computational auditory scene analysis* (CASA) has been conducted to understand the effects of a mixture of sound. However, one of the critical problems in applying CASA techniques to a real-world system is a lack of real-time processing. Usually, people hear a mixture of sounds. People with normal hearing can separate sounds from the mixture and focus on a particular voice or sound in a noisy environment. This capability is known as the *cocktail party effect*. Real-time processing is essential to incorporate cocktail party effect into a robot.

To demonstrate the feasibility of a real-time auditory and visual multiple-tracking system, it was installed in a receptionist robot and a companion robot. The system is composed of face identification, speech separation, automatic speech recognition, speech synthesis, dialog control, and the auditory and visual tracking.

## Human–Robot Interaction

Socially intelligent robots provide both a natural human–machine interface and a mechanism for bootstrapping more complex behavior. However, social skills often require complex perceptual, motor, and cognitive abilities. Current research has focused on a developmental approach to building socially intelligent robots that utilize natural human social cues to interact with and learn from human caretakers. One necessary subsystem for social intelligence is an *attention system*.

To provide a basis for more complex social behaviors, an attention system must direct limited computational resources and select among potential behaviors by combining perceptions from a variety of modalities with the existing motivational and behavioral state of the robot. Figure 2.4 shows a robotic implementation of an attention system based on models of human attention and visual search. This model interacts with existing perceptual, motor, motivational, and behavioral systems. It is the SIG humanoid developed on the Kitano Project at Kyoto University.

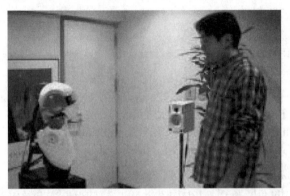

■ **Figure 2.4**   *SIG/human conversation.*

Built-in sensors enable the robotic pet to respond to both contact and a user's voice. When spoken to or touched, the robot reacts with either motion or speech. In addition, the robot can store data on interaction with its users. This information can then be accessed remotely.

## Interactive Conversation

The robotic pet shown in Figure 2.4 can contribute to relieving stress in the lives of senior citizens who live alone, by serving as a companion. The robot can also help ensure the safety of the elderly who live alone, without intruding on their privacy.

The input mode consists of sets of speech recognition and sensors, while the output mode consists of sets of speech, facial expression, and hand, leg, head, and ear (cat-type only) motions. Various information is processed to enable the robot to respond instantly to its user's actions.

Autonomous robot technology activates the robot automatically and initiates dialog with its users. The robot features a built-in, real-time clock chip. This memory chip stores recent information logs, which tell when, what, and how the robot interacted with its user. Logs are analyzed according to the laws of probability, allowing the robot to decide for itself when to wake up and initiate dialog.

Digital communication technology allows memory stored inside the robot to be remotely accessed at a high transfer rate. The robot can communicate with various external info devices, including PCs, at more than 64Kbps.

Through communication protocols specifically designed for the robot, its behavior and dialog content can be quickly updated to stimulate daily conversation, in much the same way a living pet would. Five phrases, including greetings, are recognized through speech recognition. In addition, speech synthesis has fifty phrases stored in memory. The multimodal dialog process, optimized for the elderly, enables the robot to have a personality with which any potential user can quickly become familiar. It delivers greetings and local news and information to provide opportunities for conversation. In the morning, the robot automatically wakes up and says "Good morning." It can also mix news and other voice information (downloaded to users from the local government through its serial port) into the dialog in a natural manner.

## Wireless Connection

MentorBots need ways of communicating with other entities through wireless connections. The connection can be with another robot, an electrical appliance, a phone service, or the Internet. Humans are familiar with wireless connections to another human, a stereo, a TV, a VCR, a cellular phone, or Palm Pilot.

These types of connections need to be available for MentorBots, so that they can act in behalf of their human masters. In some cases, it is only a need for convenience and in others, it may be for a life-saving emergency. In any case, it is to the benefit of humans for the robots to have such capabilities.

## Visualization

A visualization model integrates evidence from current research to construct a flexible model of human visual search behavior. In this model, visual stimuli are filtered by broadly tuned "categorical" channels (such as color and orientation) to produce feature maps with activation based on both local regions (bottom-up) and task demands (top-down). The feature maps are combined by a weighted sum to produce an activation map. Limited cognitive and motor resources are distributed in order of decreasing activation. This model has been tested in simulation, and it yields results that are similar to those observed in human subjects. No attempt is made here to match human performance (a task that is difficult with current component technology), but rather to require only that the robotic system perform enough like a human that it is capable of maintaining a normal social interaction.

A CCD-based implementation is similar to other models based in part on past researcher, but operates in conjunction with motivational and behavioral models, uses moving cameras, and differs in dealing with habituation issues.

### Identifying Surroundings

Humans receive a large amount of their information through the human vision system, since it enables them to adapt quickly to changes in their environment. An intelligent machine, such as a mobile robot that must adapt to the changes of its environment, must also be equipped with a vision system, so that it can collect visual information and use this information to adapt to its environment. In a nationwide competition to build a small, unmanned autonomous ground vehicle that can navigate around an outdoor obstacle course, the major components of the vehicle are: the supervisor control computer, speed control, steering control, obstacle avoidance, braking system, emergency controls, and vision system. The purpose of the vision system is to obtain information from the changing environment—the obstacle course. The robot then quickly adapts to this information through its controller that guides the robot to follow the obstacle course. Modeling of vision system is done with a CCD camera.

The obstacle course that the robot is supposed to follow is bounded by solid and dashed lines 10 feet apart, and can assume different shapes. Via the medium of the CCD camera, the lines of the obstacle course are digitized by the vision system from a 3-D coordinate system to a 2-D coordinate system. An image-processing tool for the vision system displays the image of the line coordinate system, reduced to 2-D coordinate system by the camera system. The robot easily obtains the information about the 2-D image coordinates from the vision system. In an autonomous situation, how can the 3-D information or coordinates of a line be determined given its image coordinate? A mathematical and geometrical transformation occurs via the camera parameters in transforming a 3-D coordinate system to a 2-D system. If these mathematical and geometrical relations are known, a 3-D coordinate point on a line can be autonomously determined from its corresponding 2-D image point. To establish these mathematical and geometrical relationships, the camera has to be calibrated. The calibration of the camera or the vision system is an important task, because if the vision system is well calibrated, accurate measurements of the point coordinates on the line with respect to the robot will be made. From these measurements, the orientation of the line with respect to the robot can be computed. With these computations, the next task is to guide the robot.

## Visual Perception

To demonstrate people awareness, the robot must be able to perceive a variety of features in real time. Several of these are visual cues. Real-time video processing is a demanding task, so an optimized generic capture and processing system capable of high throughput and low latency is needed. This specialized hardware offers significant performance enhancement for the type of arithmetic necessary for processing image data streams. A collection of visual feature detectors is under development (e.g., skin tone, motion) as are various visual behaviors (e.g., orientation, motion tracking, and target selection). An intelligent scheduler will decide the appropriate visual algorithms to use at a given time. An attention system will determine what features should be actively searched for and attended to.

The visual processing must detect several faces; extract, identify, and track each face simultaneously; and recognize the size, direction, and brightness of each face as it changes frequently. The key idea of this task is the combination of skin-color extraction, correlation-based matching, and multiple-scale image generation.

The face identification module projects each extracted face into the discrimination space and calculates its distance to each registered face. Since this distance depends on the degree (L, the number of registered faces) of discrimination space, it is converted to a parameter-independent probability.

The discrimination matrix is created in advance or on demand by using a set of variations of the face with an ID (name). This analysis is done using *online linear discriminate analysis* (OLDA). The face localization module converts a face position in a 2-D image plane into a 3-D world space. Suppose that a face is $w \times w$ pixels located in (x,y) in the image plane, whose width and height are X and Y, respectively. Then the face position in the world space is obtained as a set of azimuth 0, elevation 0, and distance r.

A real-time auditory and visual tracking system of multiple objects is needed for a MentorBot to perform its duties. Real-time processing is crucial for sensorimotor tasks in tracking, and multiple-object tracking is crucial for real-world applications. Multiple sound source tracking needs the perception of a mixture of sounds and the cancellation of motor noises caused by body movements. Real-time tracking is attained by fusing information obtained by sound source localization, multiple-face recognition, speaker tracking, focus of attention control, and motor control. Auditory streams, from 48 KHz sampling sounds, with sound source direction are extracted using an active audition system with motor noise cancellation capability. Visual streams with face ID and 3-D-position are extracted by combining skin-color extraction, correlation-based matching, and multiple-scale image generation from a single camera. These auditory and visual streams are associated by comparing the spatial location, and associated streams are used to control focus of attention. Auditory, visual, and association processing are performed asynchronously on different PCs connected by TCP/IP network.

For each of these abilities, several kinds of models exist, for example within control theory, pattern recognition, animal learning theory, and cognitive science. The goal is to develop these models in a way that makes it possible to combine them in a unified system.

Spatial learning has traditionally been tackled with the same learning mechanisms as other areas of AI and robotics, without much concern for the special requirements of this domain. As with many other techniques within AI, sensory learning and motor control have been considered problems distinct from map learning and path planning.

The traditional architecture of autonomous robots can be divided into three modules, each with its set of problems: perceptual categorization, planning and reasoning engine, and execution interface to motor functions.

A modularization of this kind is usually referred to as a *horizontal decomposition* of the robot. In brief, the main problems with this approach are:

- The computational complexity
- The interface between the plan and motor control
- The delayed feedback caused by the complexity of perception and planning
- The instability of the locomotion control as a consequence of delayed feedback

These problems have been approached during recent years by moving away from the traditional architecture in different ways. Our view of spatial orientation and learning have much in common with these newer research directions.

## Behavior-Based Reaction

A number of investigations have shown that it is possible to attack the problem of spatial learning and motor control in a different way. We refer to what is usually called a *vertical decomposition* of the system, in which the whole chain, from sensory signals to motor control, is considered continuously. These studies suggest new ways of controlling autonomous robots based on a close coupling between sensors and effectors, which can avoid the problems outlined earlier. According to the alternative principles, the construction of a system should progress from simple connections between sensors and effectors that control fundamental actions, such as moving forward without colliding with obstacles, towards more complex behaviors that may be controlled by global maps of the environment. The alternative architecture emphasizes reactive control and making the path from sensors to effectors as short as possible (Figure 2.5).

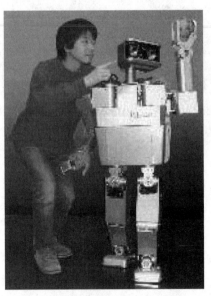

■ **Figure 2.5**  *Reactive behavior.*

If the map controlling the robot is constructed using the actual sensory and locomotion equipment of the robot, it is possible to construct plans that can be executed reactively in a stable manner. This has the advantages of the reactive approach, in that sensory signals are almost directly converted into motor commands.

Purely reactive control of behavior, as well as exploratory behavior based on simple sensory systems such as IR sensors and sonar sensors, had been intensively studied recently. Some systems using more complex inputs, such as a laser scanners and vision, also exist. These systems are not very computationally demanding, since they build on certain invariants of the environment. This feature makes them cheap to manufacture and hence attractive for practical applications.

So far, none of these systems uses visual input to learn global maps of its environment, probably because vision has traditionally been considered a very computationally demanding process. One of our main goals is to extend the reactive types of robot architectures with spatial learning abilities based on visual input.

## Cognition and Behavior

The reactive approach lends itself naturally to a view of cognition as a hierarchical adaptive control process. The view that behavior should be based on control theoretical notions, and not on planning and deduction, was pioneered by W. T. Powers and was repeated by A. H. Klopf, J. S. Morgan, and S. E. Weaver. In the spatial domain, the goal of the robot is to achieve a certain value for its spatial location. Its current location is considered a deviation from this desired value.

Research in neurobiology has provided evidence that emotions pervade human intelligence at many levels, being inseparable from cognition. Perception, attention, memory, learning, decision-making, social interaction, and communication are some of the aspects influenced by emotions. Their role in adaptation has likewise been evidenced by these studies. In the AI community, the need to overcome the traditional view that opposes rational cognition to absurd emotion has also been acknowledged. Emotion is no longer regarded as an undesirable consequence of our embodiment that must be neglected, but as a necessary component of intelligent behavior that offers a rich potential for the design of artificial systems and for enhancing our interactions with them.

This view of spatial orientation makes it possible to construct, in a unified manner, control mechanisms that combine sensory processing and motor control with spatial learning. Analysis of the constructed maps in terms of stability and optimality can also be made in a direct way while still retaining a more classical type of analysis in terms of soundness and completeness.

## Potential Fields Methods

Another means to achieve stable locomotion control is by using the potential fields methods. In this approach, goals and obstacles in the environment are given positive or negative potentials that generate gradient fields in space. By following these gradients, a robot will reach a specific goal object without running into walls and other obstacles.

The representation of the environment as a potential field, and more generally as vector fields, is a powerful way to understand the spatial representations constructed by an autonomous robot. It directly addresses two problems of the traditional approaches (i.e., the interface between the plan and motor control and the stability of the control scheme). We use potential fields as way to globally analyze the behavior of a robot while keeping the local analysis in a robot-centered representation.

## Visual Place Detectors

Vision is the only sensory system that works at all distances while combining distance sensitivity and object recognition. Tuned visual detectors are place-recognition devices that can be taught to generate a local generalization surface around any goal point in an environment. In a region around the goal point, the output of a tuned detector generates a stable control strategy for the approach of the goal. We have developed a new type of visual place detector, whose output can be tuned to produce a maximal response at any location in space. By moving the robot towards the maximum of this output, it can be made to locally approach any location in space.

The visual algorithm is based on a new type of unsupervised neural network that can associate between the visual input and a corresponding place category, as well as to similar visual views. As a consequence, the network forms an adjacency net of visual views. This approach to visual representation is very different from the traditional view, since it does not try to construct an object-centered representation of the visual scene. The network design is inspired by a number of neuropsychological findings.

The algorithm is quite fast and does not require any complex and time consuming visual preprocessing, such as segmentation or object recognition. Thus, our analysis of the visual scene is very shallow compared to other approaches, but is sufficient for spatial navigation. This part of our work has progressed to a point where a demonstration of the algorithm in an unknown environment is, in principle, already possible.

## Mapping of Spatial Locations

Tuned visual detectors can successfully control the approach to an arbitrary goal location. But since this method applies only to local regions around the goal, we intend to extend the mapping process to the entire environment. To do this, the whole environment is mapped, using tuned detectors, into a set of approach regions that cover the entire space. It is necessary to cover the environment with a large number of these regions. The regions are then linked together in such a way that the goal can be reached via a succession of subgoal locations.

We have developed a learning method that can be used to link locations in space into maps of the environment based solely on unanalyzed sensory information and the locomotion repertoire of the robot. In this work, however, the sensory system was based on simulated "olfaction" and not on vision. We will continue to develop new architectures for neural networks for the dynamic updating of spatial maps.

One goal is to combine this method with the tuned detectors described above in order to implement the whole chain from visual signals to locomotion control. We are currently evaluating the mapping system using ultrasonic sensors and active ultrasonic landmarks. In the second step, visual detectors will be used for this task.

This mapping technique is based on *reinforcement learning*. The whole process can be viewed as stochastic learning automata that establish a goal gradient for the environment. *Goal gradients* are similar to plans in the traditional approaches, except that they include a stable control strategy and present mechanisms that can be used to select dynamically between a number of goal gradients depending on the current goal of the robot. During learning and exploration, the reactive strategies of the robot play a role similar to search heuristics in the traditional approaches. If the autonomous exploratory behavior of the robot is replaced by explicit manual control, the robot can learn by instruction as well as by exploration. In this case, the automatic mapping features are only used to keep track of changes in the environment.

Reinforcement learning is usually considered too slow for use in path learning. To achieve efficiency, it must be complemented with some pre-processing, in the form of place recognition or establishment of location in space. This learning algorithm shows many similarities with *Q-learning*, but is much faster since it exploits a number of aspects of the spatial domain as well as using a reactive control system as a search heuristic. This learning method has also been thoroughly studied in computer simulations.

The mapping process is highly dependent on the attention system and the exploratory behavior used by the robot. The study of such behaviors is thus a central goal.

## MentorBot Design Considerations

In MentorBot design, the goal is to develop a scientific understanding of how people interact with an autonomous and task-oriented, but necessarily social, robot. A key issue in this work is how people develop a mental model of a robot as a machine with humanlike attributes. We build on this work to create principles for the design of a robotic assistant—the form factors, dialog, interaction routines, and feedback mechanisms that meet social as well as practical goals that encourage cooperation but not inappropriate dependence.

## Architecture

The extent to which human joint function can be replicated, along with the resulting degrees of freedom (DoF), is a key factor in robot design. In cases of multiple degrees of freedom (for example, the hip), robotic joints are implemented sequentially through short links rather than as spherical joints. Other key differences from the human form are the lack of a continuous flexible spine, and the lack of a yaw axis in the ankle. Another point to note is that the roll and pitch axes of the ankle are orthogonal, whereas the human ankle has an angle of about 64° between the roll and pitch axes.

Humanoid robots are arguably well suited to social situations. Sharing a similar morphology, they can communicate in a manner that supports the natural communication modalities of humans through facial expression, body posture, gesture, gaze direction, and voice.

## Head and Face Choices

The head and face are most important in making a MentorBot appear human. Figure 2.6 shows four examples of the important features—eyes, ears, mouth, and the shape of the head.

■ **Figure 2.6**  *Heads and faces.*

It might be assumed that all humanoid robots have facial features, but this is not the case. However, the presence of facial features is very important and figures greatly in our perception of humanness in robot heads. The three features that most increase the perception of humanness are eyes, nose, and mouth.

## Torso Choices

Because a MentorBot must be upright to get eye-to-eye with a child or seated adult, several torso designs can be considered.

Figure 2.7 shows four upright body styles. The upper left design is a simple cylindrical shape with openings or ports for access or attachments; it is the most economical to make. The upper right torso uses an LCD display as part of the structure, and it can double as the head and face. The lower pair are advanced torsos for bipedal MentorBots.

■ **Figure 2.7** *Torsos.*

## Arm Choices

Although working arms are not an absolute necessity for a MentorBot to function, they are useful. It is important for future social robots to do tasks and behave more complexly by using arms, voice, and so on. Arms may aid communication between people and robots, and these arms may be needed for aiding the elderly.

Figure 2.8 shows that a variety of arms have been developed to produce organic movement for robots using degrees of freedom. The main observed movement will be expressive in nature rather than target based.

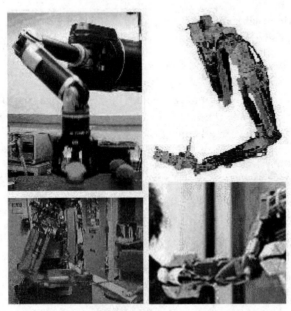

■ **Figure 2.8**  *Robot arms.*

## Locomotion Choices

It is not necessary for a MentorBot to move to perform its duties. A number have been built that are stationary until moved by the user.

Certain MentorBot tasks do require them to move, so Figure 2.9 shows the five primary methods of moving MentorBots: two-wheel drive, six-wheel drive, twin tracks, four legs, and two legs. The most common and affordable for indoor use is the two-wheel drive with one or two casters for stability. For outdoor or uneven surfaces, one of the other four choices should be used.

## Mechanical Design

One critical aspect of giving a robot a lifelike presence is designing it to move in an organic manner. The mechanical design must address the

**■ Figure 2.9** *Locomotion choices.*

control challenges of being able to position the end-effectors accurately to move smoothly and gracefully (but also quickly at times) for a broad expressive range.

Locomotion involves most of the mechanical hardware of the robot: the motors, chassis, and wheels. One of the main challenges will be figuring out how to control the motors accurately. A mobile robot must, by definition, move, and the quality of its locomotive system is key for the overall performance of the robot.

## Bipedal Walking Humanoid

There are several reasons to build a robot with humanoid form. It has been argued that to build a machine with humanlike intelligence, it must be embodied in a human-like body. Others argue that for humans to interact naturally with a robot, it will be easier if that robot has humanoid form. A third, and perhaps more concrete, reason for building a humanoid robot is to develop a machine that interacts naturally with human spaces. The architectural constraints on our working and living environments are based on the form and dimensions of the human body. Consider the design of stairs, cupboards, and chairs; the dimensions of doorways, corridors, and benches. A robot that lives and works with humans in an unmodified environment must have a form that can function with everyday objects. The only form that is guaranteed to work in all cases is the humanoid form.

The mechanical design of the humanoid requires careful and complex tradeoffs between form, function, power, weight, cost, and manufacturability. For example, in terms of form, the robot should conform to the proportions of a 4 foot-tall human. However, retaining the exact proportions compromises the design in terms of the selection of actuation and mechanical power transmission systems. Affordable motors that conform to these dimensional restrictions have insufficient power for the robot to walk or crouch.

## Animation

Researchers are exploring a variety of ways to produce organic movement for robots with many degrees of freedom. A significant portion of the observed movement will be expressive in nature (e.g., a fear response) rather than target based (e.g., orienting to a stimulus). Two methods of scripting animations are under development. The first approach uses an animation package (3D Studio Max) to script animations for a graphical model of the robot. The second approach uses a robot telemetry device to perform motion capture.

## Touch

Although robot hands emulate the structure of human hands, they are far from dexterous. Indeed, traditional robotic grippers lack flexibility. Their fixed-jaw geometry cannot grasp parts with changing orientation and securely hold them.

The solution to the problem lies in *haptic* devices. The term haptic is based on the Greek word *haptikos*, meaning to grasp or perceive. Researchers are developing tactile sensing devices that allow robots to touch things and actually feel them. Special sensors attached to the end of a robot arm can detect whether the machine is touching something soft, such as a sponge, or something hard, such as a round metal part, and apply the appropriate force.

Haptic applications for industrial robots have taken longer to evolve than machine vision, because it is a more difficult challenge. According to Ralph Hollis, Ph.D., principal research scientist at the Robotics Institute of Carnegie Mellon University, rendering visual images is a one-way street, whereas haptic devices are two-way. "Eyes take in photons, but don't shoot them out," explains Hollis. "A hand manipulates, but there is force feedback, too. So, any kind of haptic device that we use to interface with robots must take input from users as well as deliver output through the same mechanism."

Tactile sensing is more complex than vision or hearing. For example, to satisfy the eye that an image is moving, it is enough to display fifteen still pictures a second. The haptic equivalent—fooling a fingertip into believing it's feeling a surface—takes a thousand impulses a second. In addition, whereas eyes respond exclusively to light, hands and fingers respond to force, vibration, and temperature.

The human touch system also is very sensitive. "There are about 2,000 receptors in each of our fingertips whose only role is to gauge qualities like texture, shape, and the ability to cause friction," says Mandayam Srinivasan, Ph.D., director of MIT's Laboratory for Human and Machine Haptics. "There may be even more sensors for gauging warmth or coolness, and for detecting mechanical, chemical, or heat stimuli."

The BioRobotics Laboratory at Harvard University is attempting to define the ways that tactile information can improve robot dexterity. Much of the research revolves around vibratory information that can signal important events, such as the first instant of contact and the onset of slip.

"These events convey information about the state of the hand-object system that is essential for robust control of manipulation," says director Robert Howe, "Vibrations also provide perceptual information about properties, such as surface texture and friction."

The Dextrous Manipulation Lab at Stanford University is using a cyber glove to interact with a two-fingered robotic arm that feeds back force. Whatever the user does, the robot does. The robot feeds back what it "feels" to flexible sensors embedded in the glove's fingers.

Despite the challenges confronting haptic researchers, Hollis believes grippers that mimic human dexterity will open new doors for industrial robots. Bill Townsend, president of Barrett Technology Inc., is converting laboratory research into real-world products. His company's BarrettHand is an intelligent, highly flexible eight-axis gripper based on haptic research from MIT and the University of Pennsylvania.

The hand reconfigures itself in real time to conform to a wide variety of parts shapes without tool-change interruptions. Each of the grasper's three humanlike fingers, capable of 180 degrees of movement, is independently controlled by one of three servomotors, which are housed in the palm body, along with microprocessors, sensors, and other electronics. The fingers grip objects by curling together in a fist and applying pressure.

## Smell

Many products, such as automobile seats and interior modules, need to be checked for unpleasant smells before and after assembly. Traditionally, this task is carried out by professional sniffers who possess an acute sense of smell.

Unfortunately, the human olfactory system is extremely subjective. Different people are affected in different ways by similar odors. And, humans are limited by the number of products they can smell in one day.

Automated systems, such as electronic noses, have been developed to supplement human sniffers. An electronic nose is a device used to analyze the content of air through the classification of odors. It uses an array of very small sensors to detect gaseous molecules and simulate the odor-sensing capabilities of the human nose.

Electronic noses have been used successfully in the food industry for applications such as checking the freshness of cheese, fish, and fruit. But, the technology is also applicable to assembly processes, especially in the auto industry.

Engineers at Ford Motor Co. are using electronic noses to identify good and bad samples of carpet, cloth, leather, plastic, wood, and other materials used in automotive interiors. The e-nose uses an array of twelve chemical sensors. Each sensor responds to different components within an aroma to produce a "fingerprint" that identifies the material under test. The machine can then seek a match with other fingerprints in its memory.

The individual sensors have a polymer surface that acts as a conductor between two electrodes. The polymers react with the aroma molecules in an air sample, varying their electrical resistance. The change in voltage across the polymer is then measured by passing a current between the two electrodes. Each of the twelve conductive polymer sensors has a different structure and responds to different molecules.

The electronic nose operates in conjunction with an autosampler, in which test materials in small glass vials are heated for 3 hours at 80° C. A needle is then automatically inserted through the rubber seal in the vial to draw off a sample of air, including aroma molecules, and feed it to the sensor head for testing.

## Taste

Compared to the senses of sight, hearing, touch, and smell, scientists know relatively little about human taste. Indeed, taste is the most mysterious human sense and the hardest to replicate with robots or sensors. It is subject to more personal nuance than vision, touch, and other senses.

Not surprisingly, robotic taste applications lag behind other sensory research. That probably is a good thing. After all, an industrial robot with a taste disorder could devour a lot of expensive parts.

Similar to the sense of smell, the sensation and sensory process behind taste is based on chemistry. A fourfold classification system is used to determine taste: sweet, salty, bitter, and sour. Researchers have tried to mimic human taste buds by linking together sensors that detect a variety of compounds.

Alpha MOS America recently unveiled what it claims is the first electronic tongue. The Astree sensing system tests liquids and conducts taste analysis. Dissolved organic and inorganic compounds are tested for qualitative and quantitative applications.

Most applications for automated taste systems center around R&D and quality control in the food, beverage, cosmetic, and pharmaceutical industries. But possible industrial applications for a robotic arm equipped with taste sensors would include testing paints, coatings, sealants, adhesives, and solders before or after they are applied to parts. Leak detection and testing is another potential application that could benefit from automated taste systems.

## Navigation

As humans, we enjoy the luxury of having an amazing computer, the brain, and thousands of sensors to help navigate and interact with the real world. This product of eons of evolution has enabled our minds to model the world around us based on the information gathered by our senses. To navigate successfully, we can make high-level navigation decisions, such as how to get from point A to point B, as well as low-level navigation decisions, such as how to pass through a doorway. The brain's capacity to adapt has also made it possible for people without certain sensory capabilities to navigate through their environments. For example, blind people can maneuver through unfamiliar areas with the aid of seeing-eye dogs or canes. Even without all our sensors, we are able to cope with familiar and unfamiliar environments.

The abilities of modern robots stand in stark contrast to human abilities. Robots have far fewer sensors and a far less sophisticated computer; consequently, they have great difficulty adapting to changes in their capabilities and environments. Common types of robotic sensors include touch sensors for collision detection and sonar sensors for detecting obstacles at a distance. Some highly sophisticated robots use stereo vision to determine information about their environment, but this requires first solving the substantial problem of computer vision. Alas, most robots "experience" their environments through rudimentary sensors, and must process the limited information these sensors provide to make low-level navigation decisions. The problems of robot navigation are intensified if certain sensor capabilities are lost due to malfunction or damage incurred during operation.

## Sensing Navigation

### Joint Sensing

Current sensing can be performed in high power joints by using a 0.01Ohm resistance in the ground leg of the H-bridge. The voltage from these sense resistors is amplified by differential amplifiers and measured by an analog to digital converter (ADC). Current is also checked against a screwdriver-adjustable hard limit that is used to trigger the power drive protect interrupt. The position feedback from the encoders on the high power joints provides a count on every edge of both quadrature channels. This provides 2000 counts per motor revolution from the 500-count encoder wheels. In addition, each digital signal processor (DSP) can measure the bus voltage, and the temperatures of the MOSFETs and motors.

### Motion Sensing

In addition to sensing in each joint (and visual feedback), a robot may feature $2 \times 2$-axis accelerometers to provide information about the torso's dynamic behavior and the relationship to the vertical gravity force. While it is impossible to resolve the motion components of the body's acceleration from the effects of gravity, these sensors may be able to provide information with regard to disturbances while walking, playing a role similar to that of the human middle ear.

Provision can also been made for contact switches in the feet and in the joints. These switches may prove useful for determining when contact is made with the ground, or initializing joints at robot startup.

## Sonar Navigation

While many mobile robots come equipped with ultrasonic range sensing (sonar), accurate map building and position estimation using sonar has been elusive because of the difficulty in interpreting sonar data correctly. However, sonar can in fact fulfill the perception role for the provision of long-term autonomous navigation in a broad class of man-made environments.

A new approach to mobile robot navigation unifies the problems of localization, obstacle detection, and map building in a combined multitarget tracking framework. The primary tools of this approach are the Kalman filter and a physically based sonar sensor model. Experimental results with real sonar data demonstrate model-based localization using an *a priori* hand-measured map and subcentimeter accuracy map building for an uncluttered office scene.

This is of particular interest to researchers in mobile robotics, especially potential users of sonar. The approach has greater significance, however, because the issues involved—the choice of representation, the problem of data association, and the pursuit of long-term autonomy—are central to many outstanding problems in robotics and artificial intelligence.

55

## Vision Navigation

The need for autonomous robots is growing as mobile robots find an increasingly large number of applications in the areas of manufacturing, hazardous materials handling and surveillance. The basic task in any such application is the perception of the environment through one or more sensors. Processing of the sensor input results in a particular representation of the environment, which can then be used for planning actions and controlling the robot. Navigation in a previously unknown environment is a quest that many vision researchers have tackled over the years.

Although significant advances have been made, many existing mobile robot systems are designed solely for path planning, obstacle avoidance, and navigation in constrained environments. They fail to represent the environment adequately for use in different applications. Many existing systems require a model of the environment to achieve the specific tasks for which they were designed. The Computer and Vision Research Center at the University of Texas at Austin has focused on the role of machine vision for the autonomous navigation of a mobile robot in indoor, structured, previously unknown environments and the construction of a CAD model of the perceived environment.

The overall goals of the research are to develop autonomous intelligent machines that can sense the environment, estimate their position in the environment by relating sensed information to *a priori* knowledge, and effectively chart out plans and execute them for a specified task. (Although our primary interest is in the role of vision for autonomous mobile robots, we also examine the appositeness of other sensing methods, such as radar and thermal sensors, for mobile robot applications.)

### Spatial Orientation

To realize an autonomous system capable of spatial learning and spatial orientation, one must solve a number of problems. During exploration and learning, the system must accomplish the following tasks:

- **Visual place categorization.** Place representations must be constructed that can be used at later stages to recognize the location of the robot. The place categories must also make it possible for the robot to approach any location in the environment.
- **Map learning.** The robot must construct structures that can later be used to guide its locomotion from one location to another. This learning should be accelerated by using earlier spatial knowledge and requires an appropriate exploratory behavior.

Our research focuses on the following abilities, which the robot must be able to perform when using the constructed map:

- **Place recognition.** The robot must be able to figure out where it is located on the basis of visual information only. This process involves the recognition of visual views or landmarks and a potential updating of the spatial map.
- **Action selection.** The robot must determine what action to perform to move closer to the goal. This mechanism is closely connected to dynamic task selection and goal prioritizing.
- **Stable approach.** The robot must be able to approach its goal from any position within a region around the optimal path. If changes in the environment or imperfections in the motor system get the robot off course, it should automatically try to approach the correct path again.
- **Reactive obstacle avoidance.** The robot must be able to avoid obstacles in a reactive manner without too much computational overhead. Once the object is negotiated, the robot should continue

on its way toward the goal. This ability rests on a combination of different sensory systems.

Unfortunately, the method is very slow for most applications since every action must be tested a large number of times at each location. This has made reinforcement learning techniques impossible to use in mobile robots, since the learning time would be too large for any practical use. However, recent progress in this field has shown that if reinforcement learning is combined with an adaptive forward model of the environment, fast learning is possible (Peng and Williams 1993). Our approach to spatial mapping is based on a combination of reinforcement learning and potential fields methods and is much faster than earlier methods, since it exploits a number of properties of the spatial domain. It is different from the potential fields methods in that it utilizes robot-centered action-based representations.

## Stochastic Learning

Reinforcement learning, and Q-learning in particular, can be used to link actions together in such a way that the execution of the action sequence results in a maximal payoff from the environment. The learning progresses by testing a set of actions in a number of situations and collecting an immediate payoff (or reward) from the environment. The sequence of actions that results after learning is the one that returns the maximal reward when it is executed. A crucial advantage of reinforcement learning, when compared to other learning approaches, is that it requires no information about the environment except for the reinforcement signal. It also combines sequential learning with optimization in a simple way.

The central goal is to develop a robot that is able to solve various problems of spatial navigation. The sensory inputs of the robot will be a combination of tactile, ultrasonic, and visual information. We strive for a robot that can solve the following problem types, in increasing order of difficulty: reactive obstacle avoidance using tactile and ultrasonic sensors; place recognition based on ultrasonic information only; exploratory behavior; visual obstacle avoidance; visual place recognition; goal-seeking behavior using both ultrasonic and visual information; attention focusing on changes in the environment; and linguistic production of information concerning such changes, using either speech synthesis or written output on a monitor. However, so that the robot is capable of performing these tasks, we are developing theoretical models for the following three areas.

## Navigation Ability

A single camera can be used to navigate through man-made structured environments using the boundaries, or edges, of the walls and doorways it identifies. To do this, the system relies on the geometrical properties of vanishing points to estimate the most likely orientation of the edges in each 2-D image. For example, vertical edges in a 3-D scene appear to converge to one point in the image, the vanishing point of vertical lines.

This approach reduces the number of unwanted features, increases the sensitivity to useful features, and drastically speeds the computation. With the accurate estimation of the edges and boundaries in an 2-D image, the system is visually able to measure distances between edges with a precision comparable to that available from an architect's plan. This information is now being used to generate complete CAD models of building corridors.

We are also studying the use of parameter estimation and signal detection theories to recover 3-D structure for mobile robot applications. The proposed vision system recovers 3-D information from two images of the 3-D world through feature extraction, feature matching, and 3-D structure computation. The feature matching stage is implemented using an algorithm based on perceptual grouping and relaxation labeling.

Recently, an algorithm has been developed that assists the robot to avoid stationary and moving obstacles. A single wide-angle lens camera is used along with a prior knowledge of the environment. Most of the obstacle avoidance algorithms in use today employ other sensors, such as radar, to supplement the visual information. The goal is to enable the robot to perceive moving and stationary obstacles using only visual information. The system uses polar mapping to simplify the segmentation of the moving object from the background. The polar mapping is performed with the *focus of expansion* (FOE) as the center. A vision-based algorithm that uses the vanishing points of segments extracted from a scene in a few 3-D orientations provides an accurate estimate of the robot orientation, and the determination of the FOE is simplified.

In the transformed space, qualitative estimates of moving obstacles are obtained by detecting the vertical motion of edges extracted in a few specified directions. Relative motion information about the obstacle is then obtained by computing the time to impact between the obstacles and robot from the radial component of the optical flow.

For precise navigation in narrow corridors, doorways, and tight spaces, we use stereo information from two cameras with fish-eye lenses. The developed algorithm estimates the 3-D position of significant features in the scene and, by estimating its relative position to the features, navigates through narrow passages and makes turns at corridor ends. Fisheye lenses are used to provide a large field of view, which images objects close to the robot and helps in making smooth transitions in the direction of motion.

Calibration is performed for the lens-camera setup, and the distortion is corrected to obtain accurate quantitative measurements. A vision-based algorithm that uses the vanishing points of extracted segments from a scene in a few 3-D orientations provides an accurate estimate of the robot orientation.

This method is used, in addition to 3-D recovery via stereo correspondence, to maintain the robot motion in a purely transitional path, as well as to remove the effects of any drifts from this path from each acquired image. Horizontal segments are used as a qualitative estimate of change in the motion, direction and correspondence of vertical segment provides for precise 3-D information about objects close to the robot. Assuming that detected linear edges in the scene are the boundaries of planar surfaces, the 3-D model of the scene is generated.

In related projects, we have developed several position estimation techniques for robots navigating in outdoor environments. For mountainous terrains with no recognizable landmarks, we use digital elevation maps of the area, information from the camera geometry, and the position and shape of the observed horizontal line in four directions to determine the robot's position. For navigation in a structured urban environment, the position estimation technique establishes correspondences between sensor images and a stored model of the roof-top edges of the buildings in the environment.

## Vision Conclusions

To better understand the limitations of a robot, imagine a situation in which you are in an unknown house with only vague directions from your current location to the kitchen. Next, you put on a hard plastic suit that completely isolates you from the outside world. Inside the suit, all you can hear is a beeping sound that beeps faster as you approach walls or other obstacles in your path. You must use the beeping and vague directions to navigate to the kitchen successfully. Chances are, after some stumbling around and banging into walls, you will be able to find your way there.

Likewise, a successful robot navigation system will "find its way to the kitchen" using a full complement of sensors. Ideally, the system would also adapt to sensor malfunctions. In fact, researchers have developed several approaches to handling sensor failure, including the use of Bayesian networks, case-based reasoning, and evolutionary learning.

## Power Source

### Battery Packs

MentorBot rechargeable batteries are of five types: sealed lead-acid is the oldest technology, and the least expensive. Nickel cadmium has twice the efficiency of lead-acid, but is three times as expensive. Nickel metal hydride is a slight improvement, but at double the price of nickel cadmium. Lithium ion is 50 percent more efficient than nickel cadmium, but over double the price. Lithium metal has double the efficiency of nickel cadmium, but at double the price.

If the requirements are not too stringent, and the budget is limited, then nickel cadmium is the natural choice. For the ultimate in performance, lithium metal is the choice.

Most robot motors are 6-, 12-, or 24-volt DC. To reach these voltages requires the bunching of batteries into battery packs. These packs are effectively paralleled to a common bus. The packs are chosen to give 60 to 240 minutes of continuous operation.

### Automatic Recharging

In a simple example of a simple robot that has three behaviors, let's say the robot has a simple behavior that monitors the overall battery level. A second behavior randomly navigates the robot around without hitting obstacles. A third, battery-charger behavior, follows a prescribed path to the battery charger and charges up the robot.

Now consider the robot when the battery charge is good. When this is the case, the roaming behavior is the top-level behavior and the robot just wanders around.

What happens when the battery monitor behavior notices the battery charge getting low? The battery charger behavior subsumes the roaming behavior, and the robot navigates to the charger to charge up. These subsummative layers can be added on top of one another to create complex behavior modifications and reactions to outside stimulus.

Battery recharging should be handled automatically. The battery has a limited life in continuous service and then requires recharging. A residual power must be maintained in the battery to locate the charging station (by means of a visual homing beacon) after the battery monitor circuits detect a low condition. The homing beacon is activated by a coded signal sent from an RF transmitter located atop the robot's head, and the recharging supply is activated only when a demand is sensed after connection. The robot could elect to seek out the recharging station before a low battery condition actually arose, such as between patrols.

## Drive Power Electronics

The drive power electronics of the MentorBot are based on a switch-mode power stage, requiring only a single supply rail and having an efficiency over 90 percent. This efficiency results in several advantages, such as small size, lower cost power devices, and less heat sinking. The H-bridge channels are driven from separate pulse width modulation (PWM) outputs of the DSP, allowing the dead band features of the PWM peripheral to be used, along with the immediate (<12ns) shutdown of these pins in the event of a fault that triggers the power drive protect interrupt (PDPInt) pin on the DSP.

A integrated solution, the SGS-Thomson L6203, was chosen for this design. This device uses low on-resistance and fast switching MOSFETs to give maximum efficiency and best control. The voltage limit of the devices is 48 V, and the total continuous RMS current limit is 4 A. This is a good match to the chosen motors and batteries. The total on-resistance of the power devices is 0.3 ohm. The cost of the device is low, compared to a discrete solution, and the volume and mass of the electronics is minimized by the choice of an integrated solution.

## Summary

It is possible to design a practical, affordable, autonomous MentorBot. The robot should be well proportioned in relation to the human form, with most of the major degrees of freedom of the human body implemented. The robot design should have a distributed control design, with processors dedicated to each of the key roles around the robot. Investigations of the CAD design using a high-fidelity simulation have shown that such a robot is capable of crouching and balancing.

Users enjoy interacting with a MentorBot and treat it as a living thing, usually a pet or young child. Kids were more engaged than adults and had responses that varied with gender and age. No one seemed to find the robot disturbing or inappropriate. We seemed, therefore, to have avoided any uncanny valleys (with the possible exception of vocalizations). It also got attention easily, fulfilling another of our design goals.

A friendly robot usually prompted subjects (older boys excepted) to touch it, mimic its motions, and speak out loud to it. With the exception of older boys, a sad, nervous, or afraid robot generally provoked a compassionate response. Older boys did not generally demonstrate this compassion but seemed to respect the robot only if it fought back when abused.

Our interactions with users also showed a potential need for future (autonomous) social robots to have a somewhat different sensory suite than current devices. For instance, we found it very helpful in creating a rich interaction to "sense" the location of bodies, faces, and even individual eyes on users. We also found it helpful to read basic facial expressions, such as smiles and frowns. This argues for a more sophisticated vision system, one focused on dealing with people. Additionally, it was essential to be aware of participants' immediate proximity and to know where the robot was being touched. This may mean the development of a better artificial skin for robots. If possessed by an autonomous robot, these types of sensing would support many of the behaviors (i.e., touching, mimicry, etc.) that users found so compelling.

Fortunately, there are some traditional robotic skills that an autonomous robot might not need. For example, there was no particular need for advanced mapping or navigation. The MentorBot goal is to be social, and that requires paying primary attention to people rather than efficient navigation to a target. In fact, people will usually come right to it; there was no need to seek them out. A robot that could pay attention to people in its field of view and had enough navigation to avoid bumping into objects would probably do quite well.

In addition to using somewhat different "senses" than traditional robots, a MentorBot demonstrates a different kind of intelligence than is traditionally associated with robots. Most standard research robots today use AI to accomplish a clearly delineated set of manipulations in the world. The robot has a task, and there is one correct answer. We see current and future social robots as using a form of AI often referred to as "perceived intelligence," which holds human experience as the metric for success.

The robot need only do something "reasonable" to be a success. In short, social robots can succeed with a variety of actions, and there are many correct answers for a given situation. Experience bears this out; a MentorBot was often perceived as acting sensibly even when suffering a serious control malfunction that left it behaving erratically. It will be challenging to build these new social capabilities into mobile robots, but humans provide a more forgiving environment than roboticists are accustomed to.

For instance, attentiveness was easily discerned just by watching the head of the robot as it followed a person's motion. This emotional state maps readily to a device's success at paying attention to a user. Similarly, a MentorBot could intuitively demonstrate a certain energy level using posture and the pace of its motion. This emotional state could have been related to the state of its batteries.

In short, by choosing emotional states reasonable to a given device and portraying that state in simple body language, interface designers may very well be able to open a relatively new channel for communicating with users. This "body-centric" technique, borrowed from robotics but potentially applicable broadly, should be a subject of further research.

**63**

# Enabling Technology

MENTORBOTS ARE DESIGNED TO BE CAPABLE AND APPEALING robots that can communicate, interact, learn from, and teach humans. In a unique collaboration, this project seamlessly merges the artistry of character, robotic technology, and AI. In its very early stages, current efforts are focused on building a highly expressive and organic body and the synthetic central nervous system (i.e., the computational, motor, and perceptual systems).

Exploring human–robot interaction requires constructing increasingly versatile and sophisticated robots. Commercial motor-driver and motion-controller packages are designed with a completely different application in mind (specifically, industrial robots with relatively small numbers of relatively powerful motors) and do not adapt well to complex interactive robots with a very large number of small motors controlling things like facial features. Some developmental robots, for instance, have sixty motors in an extremely small volume of space. An enormous rack of industrial motion controllers would not be a practical means of controlling the robot; an embedded solution designed for this sort of application is required.

## Conscious Thinking

The term artificial intelligence (AI) was coined at the groundbreaking Dartmouth conference of 1956. But man's interest in the notion that a machine could be given the ability to think can be traced back to the myths and stories of the ancient world.

In centuries past, myriad philosophers actively pondered some of the same questions addressed by present-day AI researchers. For example, the great French thinker Descartes—whose statement, "I think, therefore I am" identified that to be intelligent is to be alive—believed that animals were no more than automatons, or self-moving machines. But the queen of France, who insisted on seeing proof that a clock could reproduce, foxed him.

## Artificial Intelligence

Artificial intelligence has been defined as

"The art of creating machines that perform functions that require intelligence when performed by people." (Ray Kurzweil, 1990, *The Age of the Intelligent Machine*, MIT Press, Cambridge, Mass.)

and as

"The study of the computations that make it possible to perceive, reason, and act." (Patrick Winston, 1992 *Artificial Intelligence*, Addison-Wesley, Reading, Mass.)

A wide variety of AI methods and applications are represented at Mitsubishi Electric Research Laboratories (MERL):

■ In the area of perception, the computer vision researchers are working on enabling computers to perform visual tasks such as recognizing faces and fingerprints and retrieving photographs based on their content.

■ In the area of reasoning, the machine learning researchers are using both statistical and symbolic methods to make computing systems that improve their performance with experience. Research at MERL on making computers understand spoken language spans both perception and reasoning.

■ Finally, the experience of the last few decades has shown that artificial intelligence is often most effective when used as an "ingredient" in other systems. In this way, the AI ingredient is being used for applications as diverse as automatic map layout and lighting design for 3-D graphics.

## The Behavior System

The behavior system organizes a robot's task-based behaviors into a coherent structure. Each behavior is viewed as a self-interested, goal-directed entity that competes with other behaviors to establish the current task. An arbitration mechanism is required to determine which behavior(s) to activate and for how long, given that the robot has sev-

66

Chapter Three

eral motivations that it must tend to and different behaviors that it can use to achieve them. The main responsibility of the behavior system is to carry out this arbitration. In particular, it addresses the issues of relevancy, coherency, persistence, and opportunism. By doing so, the robot is able to behave in a sensible manner in a complex and dynamic environment.

### Task-Learning Architecture Using Behaviors

It is now commonly believed that intelligent agents are best implemented via multiple behaviors. For example, in the context of mobile robots, the various behaviors may correspond to "stay in the middle of the hallway," "avoid obstacles," or "seek the goal." It is not difficult to see that in any tangible application, the various behaviors will produce a discordant set of commands for the robot. The problem then becomes one of arbitrating between the commands dictated by the various behaviors and choosing one of the commands or some combination of them. This problem is known as the *problem of arbitration*. In the past, various solutions have been suggested for solving this problem, such as the *vector field approach*, the *subsumption approach*, and the *task-based approach*.

Current research presents a different solution. Behaviors are evaluated by mapping functions from the domain, which consists of sensory readings, to the co-domain, which consists of multiple sets of command decisions. In this approach, taking a weighted sum of these functions combines the behaviors. The individual weights are initially specified, but dynamically modified and learned by a reinforcement-learning scheme. All these behaviors are programmed within a framework, which is another aspect of this research. The results, both from simulation and real-world situations, support claims of solving the problem of arbitration while providing the system with the ability to adapt dynamically to hitherto unknown environments.

The mechanical sophistication of a full-fledged humanoid body poses a devastating challenge to even the most robust learning technique. The more complex a humanoid body, the harder it is to place the constraints necessary for productive learning. If too few constraints are employed, learning becomes intractable. Too many constraints, on the other hand, limit the ability of learning to scale. Ultimately, the conventional learning techniques described here are limited by the fact that they are tools wielded by human designers, rather than self-directed capabilities of the robot. This may not need to be the case. Although robots will always require an initial program, this fact does not preclude them from build-

**67**

ing indefinitely, willfully, and creatively upon it. After all, humans also begin with a program encoded in our DNA. The key is that, in humans, the majority of this genetic code is devoted not to mere behavior, but to laying a foundation necessary for future development.

## Active Collaboration

Collaborative interaction is a technology area that uses advanced computer hardware and software to help people collaborate with each other. MERL is pursuing many different approaches to improving collaborative interaction.

- MERL is designing new interaction mechanisms so that people and computers can cooperatively solve problems in such a way that people perform the roles they are best at (i.e., those requiring intelligence, learning, strategy, etc.), and computers can perform the roles they are best at (i.e., those requiring brute force computation or search). Systems are being developed in which people and computers can solve problems more effectively than either can do by themselves.

- MERL is exploring ways that computation can support face-to-face collaboration and communication between people, either to solve problems or simply to share experiences. Similarly, ways are being explored in which people can interact with computers in more "natural" ways than are afforded by the traditional interface devices such as keyboards, mice, and monitors. The goal is to help computers "disappear," as have previous successful technologies, such as writing and electricity, so that people can use computers easily without being distracted by the technology itself.

## The Motivation System

The motivation system consists of the robot's basic "drives" and "emotions." The "drives" represent the basic "needs" of the robot and are modeled as simple homeostatic regulation mechanisms. When the needs of the robot are being adequately met, the intensity level of each "drive" is within a desired regime. However, as the intensity level moves farther away from the homeostatic regime, the robot becomes more strongly motivated to engage in behaviors that restore that "drive." Hence the "drives" largely establish the robot's own agenda, and play a significant role in determining which behavior(s) the robot activates at any one time.

The "emotions" are modeled from a functional perspective. Based on simple appraisals of the benefit or detriment of a given stimulus, the robot evokes positive emotive responses that serve to bring it closer, or negative emotive responses in order to withdraw. A distinct emotive response exits for each class of eliciting conditions. Currently, six basic emotions are modeled that give the robot synthetic analogs of anger, disgust, fear, joy, sorrow, and surprise. There are also arousal-based responses that correspond to interest, calm, and boredom that are modeled in a similar way. The expression of emotive responses promotes empathy from the caregiver and plays an important role in regulating social interaction with the human.

## Social Amplification

The target MentorBot is designed to make use of human social protocol for various purposes. One such purpose is to make life easier for its vision system. If a person is visible, but is too distant for his face to be imaged at adequate resolution, the target MentorBot engages in a calling behavior to summon the person closer. People who come too close to the robot also cause difficulties for the camera's narrow field of view, since only a small part of a face may be visible. In this circumstance, a withdrawal response is invoked, where the target MentorBot draws back physically from the person. This behavior, by itself, aids the cameras somewhat by increasing the distance between the target MentorBot and the human. But the behavior can have a secondary and greater effect through social amplification—for a human close to the target MentorBot, a withdrawal response is a strong social cue to back away, since it is analogous to the human response to invasions of "personal space."

Similar kinds of behavior can be used to support the visual perception of objects. If an object is too close, the target MentorBot can lean away from it; if it is too far away, the target MentorBot can crane its neck towards it. Again, in a social context, such actions have power beyond their immediate physical consequences. A human, reading intent into the robot's actions, may amplify those actions. For example, neck-craning towards a toy may be interpreted as interest in that toy, resulting in the human's bringing the toy closer to the robot. Another limitation of the visual system is how quickly it can track moving objects. If objects or people move at excessive speeds, the target MentorBot has difficulty tracking them continuously. To bias people away from excessively boisterous behavior in their own movements or in the movement of objects they manipulate, the target MentorBot shows irritation when its tracker is at the limits of its ability. These limits are either physical (the maximum rate at which the eyes and neck move) or computational.

## Envelope Displays

Such regulatory mechanisms play roles in more complex social interactions, such as conversational turn-taking. Here control of gaze direction is important for regulating conversation rate. In general, people are likely to glance aside when they begin their turn, and make eye contact when they are prepared to relinquish their turn and await a response. People tend to raise their brows when listening or waiting for another to speak. Blinks occur most frequently at the end of an utterance. These envelope displays and other cues allow the target MentorBot to influence the flow of conversation to the advantage of its auditory processing. The visual–motor system can also be driven by the requirements of a nominally unrelated sensory modality, just as behaviors that seem completely orthogonal to vision (such as ear-wiggling during the call behavior to attract a person's attention) are nevertheless recruited for the purposes of regulation.

## Other Regulatory Displays

Some regulatory displays also help protect the robot. Objects that suddenly appear close to the robot trigger a looming reflex, causing the robot to withdraw quickly and appear startled. If the event is repeated, the response quickly habituates and the robot simply appears annoyed, since its best strategy for ending these repetitions is to signal clearly that they are undesirable.

Similarly, rapidly moving objects close to the robot are threatening and trigger an escape response. These mechanisms are all designed to elicit natural and intuitive responses from humans, without any special training. But even without these carefully crafted mechanisms, it is often clear to a human when the target MentorBot's perception is failing, and what corrective action would help, because the robot's perception is reflected in behavior in a familiar way. Inferences made based on our human preconceptions are actually likely to work.

The target MentorBot is an autonomous robot designed for social interactions with humans. In general, social robotics has concentrated on groups of robots performing behaviors such as flocking, foraging, or dispersion, or on paired robot–robot interactions such as imitation. This project focuses not on robot–robot interactions, but rather on the construction of robots that engage in meaningful social exchanges with humans. By doing so, it is possible to have a socially sophisticated human assist the robot in acquiring more sophisticated communication skills and helping it learn the meaning these acts have for others. This approach is inspired by the way infants learn to communicate with adults; the mode of social interaction is that of a caretaker–infant dyad, in which a human acts as caretaker for the robot.

The system architecture (Figure 3.1) consists of six subsystems: the *low-level feature extraction system*, the *high-level perception system*, the *attention system*, the *motivation system*, the *behavior system*, and the *motor system*. The low-level feature extraction system extracts sensor-based features from the world, and the high-level perceptual system encapsulates these features into percepts that can influence behavior, motivation, and motor processes. The attention system determines what the most salient and relevant stimulus of the environment is at any time so that the robot can organize its behavior accordingly. The motivation system regulates and maintains the robot's state of "well-being" in the form of homeostatic regulation processes and emotive responses. The behavior system implements and arbitrates between competing behaviors. The winning behavior defines the current task of the robot.

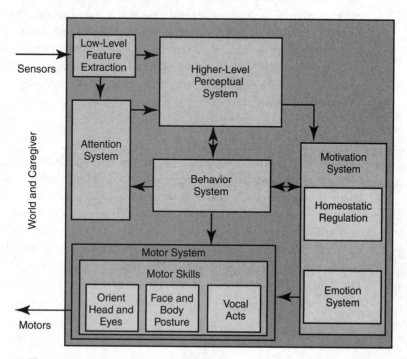

■ **Figure 3.1** *MentorBot architecture.*

The robot has many behaviors in its repertoire, and several motivations to satiate, so its goals vary over time. The motor system carries out these goals by orchestrating the output modalities (actuator or vocal) to achieve them. For the target MentorBot, these actions are realized as motor skills that accomplish the task physically, or expressive motor acts that accomplish the task via social signals.

### The Low-Level Feature Extraction System

The low-level feature extraction system is responsible for processing the raw sensory information into quantities that have behavioral significance for the robot. The routines are designed to be cheap, fast, and just adequate. Of particular interest are those perceptual cues that infants seem to rely on—visual and auditory cues, such as detecting eyes and the recognition of vocal affect, are important for infants.

### The Attention System

The low-level visual percepts are sent to the attention system. The purpose of the attention system is to pick out low-level perceptual stimuli that are particularly salient or relevant at that time, and to direct the robot's attention and gaze toward them. This provides the robot with a locus of attention that it can use to organize its behavior. A perceptual stimulus may be salient for several reasons. It may capture the robot's attention because of its sudden appearance, or perhaps due to its sudden change. It may stand out because of its inherent saliency, as a red ball may stand out from the background. Or perhaps its quality has special behavioral significance for the robot, such as being a typical indication of danger.

### The Perceptual System

The low-level features corresponding to the target stimuli of the attention system are fed into the perceptual system. Here they are encapsulated into behaviorally relevant percepts. To elicit processes in these systems environmentally, each behavior and emotive response has an associated *releaser*. A releaser can be viewed as a collection of feature detectors that are minimally necessary to identify a particular object or event of behavioral significance. The function of the releasers is to ascertain if all environmental conditions are right for the response to be active.

# Communication

## Auditory System

The caregiver can influence the robot's behavior through speech by wearing an unobtrusive wireless microphone. This auditory signal is fed into a 500-MHz PC running Linux. The real-time, low-level speech processing and recognition software was developed at MIT by the Spoken Language Systems Group. These auditory features are sent to a dual 450-MHz PC running NT. The NT machine processes these features in real time to recognize the spoken affective intent of the caregiver.

The target MentorBot has a 15-DoF face that displays a wide assortment of facial expressions to mirror its "emotional" state and to serve other communicative purposes. Each ear has two degrees of freedom that allows the target MentorBot to perk its ears in an interested fashion, or fold them back in a manner reminiscent of an angry animal. Each eyebrow can lower and furrow in frustration, elevate upwards for surprise, or slant the inner corner of the brow upwards for sadness. Each eyelid can open and close independently, allowing the robot to wink an eye or blink both. The robot has four lip actuators, one at each corner of the mouth, that can be curled upwards for a smile or downwards for a frown. There is a single-degree-of-freedom jaw.

## Spoken Language Interface

Speech recognition and speech synthesis technologies have been available in crude form for nearly two decades. However, because of technical limitations, their application has been limited to a few success stories. Meanwhile, over the same two decades, a revolution in consumer electronics and computing devices has dramatically increased the market need for spoken language interfaces to simplify UI and free up the hands and eyes. So, while speech recognition and synthesis technologies continue to make incremental improvements, the potential for a revolution in spoken language interfaces looks promising (Figure 3.2).

At MERL, the approach to research in spoken language interfaces takes two directions:

- **Speech-centric devices.** Most applications of speech technology take the approach of layering a speech interface on top of existing hardware, providing the user the option of pressing buttons with either the finger or the voice. At MERL, we are interested in creating new devices designed with spoken language interfaces in mind.

We believe that the result will be improvements in industrial design, hardware costs, and interface accuracy.

- **Conversational speech interfaces.** The poor accuracy of even the best speech recognition programs has made it impossible to consider using speech as the primary interface for complex tasks. Ironically, complex tasks most need a speech interface. At MERL, we are applying principles from human collaborative discourse theory to build a conversational framework on top of which we can layer speech interfaces. With this framework, we are able to build into the software system some understanding of the task at hand. We believe that this will make conversational, natural spoken language interfaces more accurate, robust, and useful for real products.

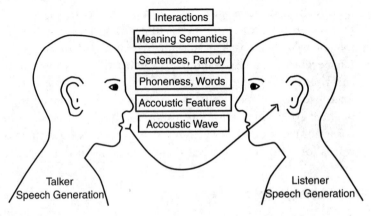

- **Figure 3.2** *Spoken language interface.*

## Vocalization System

The robot's vocalization capabilities are generated through an articulator synthesizer. The underlying software (DECtalk v4.5) is based on the Klatt synthesizer, which models the physiological characteristics of the human articulator tract. By adjusting the parameters of the synthesizer, it is possible to convey speaker personality (the target MentorBot sounds like a child) as well as adding emotional qualities to synthesized speech.

## Speech Analysis and Synthesis

Voice analysis and synthesis functions were added to the MentorBot project to make the robot more approachable and interesting. The goal of the project team was to have a robot that could understand various questions and respond accordingly. Some of the design goals were as follows:

- The speech analysis would be performed at real-time rates on continuous speech.
- No training of the robot's voice analysis functions would be required. This would allow anyone to approach the robot and initiate a conversation.
- None of the questions were preprogrammed. Rather, the analysis of the content of a question would be used to search for the best possible response.
- The voice synthesis would have the ability to change personalities by varying the speaking speed or pitch. This would give the robot a little more unpredictability.

The speech system for MentorBot had four major components: the speech analysis, speech synthesis, LISP interpreter and program, and the user interface. Each one of these components used either off-the-shelf software or modifications to public domain source code. The largest task by far was the integration of the components.

## Speech Analysis

IBM ViaVoice Gold was the program selected for the speech analysis because a member of the project team had an old version of the software. The performance of the program with a properly trained user profile was very impressive when the user was *reading* the material to be analyzed. When the material was typically conversation, performance was drastically reduced, probably because pronunciation is better and the speaker's speed has a more even cadence when reading. All training is done by reading.

Program performance was even further reduced with very short sentences that were read or spontaneous ones with one- or two-word bursts.

Presently, the robot uses a headset that helps eliminate feedback from the speakers because the microphone and speakers are essentially isolated from each other. The headset was used during the training session to improve the accuracy of speech-to-text conversion, which is partially dependent on the electrical characteristics of the microphone being used. Since training of the robot voices was done with a particular headset, the same headset needed to used the during normal conversations with the robot in order for the robot to maintain its speech-to-text conversion accuracy.

### Speech Synthesis

Various text-to-speech synthesis programs are available. The voice quality ranges from poor to spectacular. The OGI Festival speech system was evaluated and had very good performance. In the end, the project team decided to use the voice synthesis built into the newer Microsoft libraries called "Microsoft Voice." The pitch and speed are variable, and the integration into the user interface was very simple.

### LISP Interpreter—Eliza

The Lispworks interpreter from Harlequin was used as the LISP interpreter (Figure 3.3). This program provided a simple communication package that was used to interface to other blocks in the system.

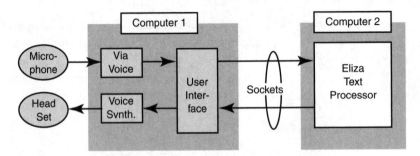

■ **Figure 3.3**   *LISP interpreter diagram.*

The program was a variation of the Eliza program developed by Matt Maple. The first thing Eliza does to an input text is strip off any punctuation and convert the text to a list. It then replaces with key words Eliza understands those words or phrases in the list that are semantically the same but syntactically different from the Eliza key words. Eliza then attempts to deduce what the individual is saying. It does this by looking for key words in specific positions relative to one another. If it finds a match, it generates the output response attached to that particular condition and stores it in the output buffer.

This principle is illustrated by the sample input text in Figure 3.3. Eliza checks to see if the individual is asking for directions to the ECE office by looking for the words "where" and "electrical and computer engineering office" in that order. In this particular example, the user could have substituted "electrical and computer engineering" with "ECE," since Eliza understands that both words mean the same thing. This form of analysis is flexible because it gives the user freedom in constructing the question.

## System Operation

All of the components described were integrated into the system using two computers. One computer was used to perform the user interface, voice analysis, and voice synthesis. The second computer performed the text-based analysis. The division was done because voice analysis consumes lots of CPU cycles. Splitting the voice processing off onto its own computer allows the LISP portion of the system to grow and control various other subsystems of the robot, such as locomotion and sensor integration. The two computers communicated via TCP/IP sockets. A rough diagram of the system is shown in Figure 3.4.

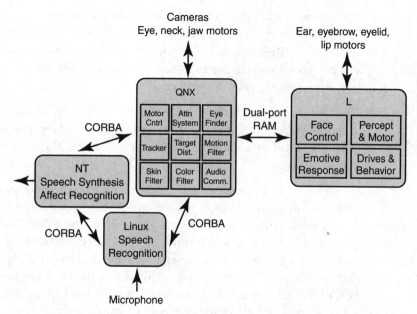

■ **Figure 3.4**  *Communication system diagram.*

The microphone captures data in the form of speech. The data is processed by the voice analysis software and sent to the user interface in the form of a text string and stored in the "input text" buffer. When the user has not spoken for 1 to 2 seconds (latch delay), the data is latched into the "transmitted text" buffer, and the user interface sends that buffer to the Eliza-like program for text processing. When the Eliza processing is completed, the response is sent to the user interface via the network and is stored in the "receive text" buffer. The user interface displays the text and also sends the text string to the voice synthesis block, which in turn produces the verbal response.

The Eliza program had the ability (or disability) to have multiple personalities by changing the vocabulary for each personality. Additionally, a system was set up whereby the Eliza program could also control the verbal characteristics of the outbound speech to match the personality. The Eliza program sends escape control words to the user interface in the normal outgoing data path. The user interface program detects the escape word and then performs the desired operation. The escape control word is not uttered. Currently only two escape commands are issued. One is for the friendly personality and the other is for the police personality. The police personality has a lower pitch and slower cadence compared to the friendly personality.

The system first developed for this project was sufficient for a well-controlled stationary environment. Unfortunately, this is not the normal operation environment for a mobile robot. The accuracy of the voice analysis portion of the system is still poor. A simpler, smaller vocabulary system would probably do better.

The conversations that the robot had with itself if the feedback from the speakers was picked up by the microphone were very amusing and usually resulted in mumbling incoherent sentences. Maybe this is the idiot personality?

In the future, the speech system needs to be made mobile and fitted onto the robot. To do this, it is necessary that the voice input be taken from a wireless camera, which is mounted onto the robot. The audio portion of this link is sent to the receiver and connected to the computer. The voice analysis computer will not be onboard the robot, mainly because of the processing required. For the voice synthesis portion of the mobile robot, a wireless link will send the audio to the robot. A modified walkie-talkie would probably suffice.

## Digital Communications

Digital communications technology has revolutionized telecommunications and computer industries over the last decade. But technology development in this field is still rich with innovation. Mobile communication is being taken for granted now, but many challenges remain, such as maintaining high quality of service in high-mobility areas, and maintaining high data transmission rates with constrained power and bandwidth.

Two overarching questions drive the research efforts in digital communications at MERL. For wireless communications, the question is, "What services, network capabilities, and long-term architecture are needed to support applications beyond IMT 2000 and 3G?"

Broadband communications raises the question, "What architectures and transmission technologies (optical, cable, wireless, power line, etc.) are required to increase the bandwidth and services offered to the user?" See Figure 3.5.

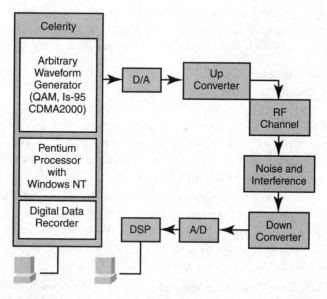

■ **Figure 3.5** *Broadband diagram.*

- **Wireless communications.** MERL is an active participant in most major industry standards for wireless communications protocol, including IEEE 802.16, 3GPP2, 3GPP SG, ITU-T, and IETF. These standards improve system architecture and algorithm design for Mitsubishi Electric products. To test the designs, a DSP-based test bed was built for FPGA prototyping.

- **10 Gigabit Ethernet.** MERL is an active participant in the IEEE 802.3ae (10 Gigabit Ethernet) standard, and is designing equalization techniques and building technology that will allow successful products based on the standard.

- **Wireless LAN.** MERL is an active participant in wireless LAN standards, including IEEE 802.11 and 1394 WLAN. MERL is also building test beds and demonstration systems, driving toward new technology for products and applications.

At MERL, the aim is to create new concepts and technologies in wired and wireless digital communications. Participation in the development of industry standards is an important part of the work. The main focus is on two areas: networks and net services.

## Networks

As computers have become smaller and more powerful, almost all devices have computing power in some form. Therefore, organizing a collection of such devices—at home, work, or mobile environments—requires the development of networking technologies that enable command and control of a collection of devices, not all of which are computers. Such technologies have to go beyond traditional computer networking techniques.

In the area of networks, MERL is exploring several different types of technologies for consumer and industrial uses:

- **UPnP.** MERL is an active participant, contributor, and proponent of universal plug and play (UpnP), an open Internet architecture for peer-to-peer network connectivity of intelligent appliances, wireless devices, and PCs of all form factors.

- **PLC.** We are developing power line communication (PLC) technology, which seeks to transmit a few kilobits to a megabit per second of data over normal AC power lines. The idea is to make use of transmission bandwidth potential available in the existing power line wire infrastructure.

- **LVDS.** Low-voltage differential signaling (LVDS) is a high-speed I/O technology for communication between silicon chips. MERL is currently monitoring developments in this high-speed technology, with an eye toward applying them to improve Mitsubishi Electric products and services.

- **DTV.** MERL works on both the broadcast and the consumer sides of networking digital television (DTV). On the broadcasting side, they support the high-definition encoders made by Mitsubishi Electric. On the consumer side, they work with DTV manufacturing groups in the U.S. on the integration of networking technologies (such as IEEE 1394, HAVi, etc.) into the DTV framework.

## Net Services

Net services technology leverages the ubiquity and accessibility of information provided by the Internet. Internet-ready devices are showing up in automobiles, hospitals, the mobile workplace, and nearly every room in the home. Soon the personal computer will not be the primary means for people to access the Internet. This makes Internet software a rich area of research and development, as the Internet continues to expand into diverse arenas such as public, private, B2B, B2C, wireless, entertainment, and location aware information services.

MERL is focused on developing middleware and applications that enable new types of Internet-based services and software products. The most significant efforts are in the areas of mobile agents and wireless Internet networking:

- **Mobile agents.** MERL has developed a mobile agent technology, called Concordia, which is currently licensed by several customers. It is a framework for developing and managing network-efficient mobile agent applications useful for accessing information anytime, anywhere and on any device supporting Java. The Concordia system allows custom mobile agents to travel across the Internet, where they can interface with back-end applications, databases, or other mobile agents. We are continuing to develop the Concordia framework, making it useful for a wider range of applications, including enterprise application Integration.
- **Wireless Internet networking.** MERL is developing a wireless Internet networking framework that allows for easy, ubiquitous access to personalized information services and content using a personal wireless communication device. This technology, applied to several MERL projects, is a cell-phone–centric networking technology that turns common consumer appliances, such as a TV set or a car navigation system, into an Internet appliance.

## Visualization

A necessary sensory aptitude for a sociable robot is to know where people are and what they are doing. Hence, the sociable robot must be able to monitor humans in the environment and interpret their activities, such as gesture-based communication. The robot must also understand aspects of the inanimate environment as well, such as how its toys behave as it plays with them. An important sensory modality for facilitating these kinds of observations is vision. The robot will need a collection of visual abilities, closely tied to the specific kind of information about the interactions it needs to extract.

### Active Vision

Towards this goal, a suite of visual capabilities was developed as the investigation continued of Intel's OpenCV library (supplementing the routines with the addition of Mac G4 AltiVec operations). This includes a collection of visual feature detectors for objects (e.g., color, shape, and motion) and people (e.g., skin tone, eye detection, and facial feature tracking), the ability to specify a target of attention and track it, and stereo depth estimation. Active vision behaviors include the ability to saccade to the locus of attention, smooth pursuit of a moving object, establishing and maintaining eye contact, and vergence to objects of varying depth.

## Vision System

The robot's vision system consists of four color CCD cameras mounted on a stereo active vision head. Two wide field of view (fov) cameras are mounted centrally and move with respect to the head. These are 0.25-inch CCD lipstick cameras with 2.2-mm lenses. They are used to decide what the robot should pay attention to, and to compute a distance estimate. A camera is also mounted within the pupil of each eye. These are 0.5-inch CCD fov cameras with an 8-mm focal length lenses, and are used for higher resolution post-attention processing, such as eye detection.

The target MentorBot has three degrees of freedom to control gaze direction and three degrees of freedom to control its neck. The degrees of freedom are driven by DC servo motors with high-resolution optical encoders for accurate position control. This gives the robot the ability to move and orient its eyes like a human, engaging in a variety of human visual behaviors. This is not only advantageous from a visual processing perspective, but humans attribute a communicative value to these eye movements as well.

## Advanced Digital Video

Advanced digital television provides an exciting new world for the consumer. Not only does it offer improved picture quality, it also provides a means for seamlessly blending many new services into the TV set, expanding the scope of what televisions can do. Advanced digital television also poses new challenges in video encoding, transmission, and reception.

Historically, the advanced digital television effort at MERL has focused on the development of DTV, HDTV, and related new technologies. The strength is in developing new algorithms and hardware for advanced digital television receivers, with emphasis on low-cost, high-quality solutions.

## Computer Vision

Computer vision combines sensors with data processing algorithms to perform "eye" functions for electronic products and systems.

Computer vision can perform functions too tedious or time consuming for humans, such as visual inspection of parts or the detection of visual anomalies (Figure 3.6). Computer vision is also a key element in developing computer interfaces that are natural and easy to use.

video input —> 3D

■ **Figure 3.6** *Video frame analysis.*

Computer vision research at MERL covers a range of technology pursuits:

- Building technology involving the recognition and interpretation of the human form, including poses and motions related to the eyes, face, head, hands, and the entire body. Applications include user interfaces for games or home appliances, access control, video content analysis, and safety systems for automobiles.

- Focusing research on low-cost sensors by developing technology that will enable new computer vision applications based on low-cost CMOS vision sensors, such as Mitsubishi Electric's artificial retina chip.

- Researching the recovery of 3-D structure from an image sequence through an integrated system that recovers 3-D models composed of points, straight line segments, or planes. The system is automatic and robust, and the goal is to make complex 3-D model generation simple—using still images or video—thus making 3-D graphics technology more commonplace.

## Audio Video Processing

Most audio video processing technology revolves around industry standards. The Motion Picture Encoding Group (MPEG) is most famous for its digital video compression standards MPEG-1 and MPEG-2. MPEG-1 targets CD-ROM-based applications, whereas MPEG-2 targets broadcast-quality video. MPEG continues to develop new standards that address the requirements of emerging systems. MPEG-4 is a video compression standard that addresses object-based processing, blending of synthetic and natural video, and low-bit rate transmission, among other things. MPEG-7 addresses the standardization of multimedia content descriptions to enable applications such as content-based remote and local browsing.

The audio video processing effort at MERL consists of MPEG-4, MPEG-7, and related activities. They have successfully proposed new technologies for adoption into these standards. The object-based rate control algorithm is now part of the informative (non-normative) part of the MPEG-4 standard. The video motion activity descriptor, directed acyclic graph-based description scheme, video transcoding description scheme, and audio content indexing and extraction method are all part of the normative part of the MPEG-7 standard committee draft. Current areas of interest include:

- **MPEG-4.** MERL is developing MPEG-4 codec technology for encoder optimization and video segmentation, and a system development effort for wireless video streaming using the MPEG-4 standard.
- **MPEG-7.** Technology for video browsing, indexing, and summarization, as well as creating new MPEG-7 enabled product and service concepts. Also, MERL is developing technology for the recognition and extraction of individual sound sources from mixed audio scenes.

## Graphic Interpretation

Most sensory information comes through the eyes. Computer graphics, therefore, plays a critical role at the interface between human and computer. MERL has a strong history in the field of computer graphics research, with participation in respected industry conferences and with diverse research projects, running the gamut from scientific visualization to interactive manipulation of geometric shapes (Figure 3.7).

■ **Figure 3.7** *Graphic imaging.*

At MERL, the key focus areas in computer graphics include:

■ **Modeling.** New methods of representing computer graphics that offer advantages for compression, transmission, editing, rendering, and animation of graphics models with increasingly complex shapes. Research includes the use of point samples and adaptively sampled distance fields as primitives for 3-D graphics.

■ **Image-based rendering (IBR).** IBR refers loosely to techniques that generate new images from other images rather than from a geometric description. It combines techniques of computer graphics and computer vision. IBR enables interactive photorealistic rendering, because the source data are photos. This may allow the creation of static and dynamic 3-D photographs and immersive 3-D environments, enabling new applications in e-commerce, entertainment, and telecollaboration.

■ **Computer graphics and machine learning.** The principles of AI and machine learning are being applied to complex computer graphics problems, such as lip synchronization for speech, computer-generated animation, and high-resolution image processing.

■ **Scientific visualization.** The large amounts of data generated by today's complex simulation and acquisition systems require sophisticated visualization techniques. MERL research in this area led to the creation of VolumePro, the world's first real-time volume rendering hardware. VolumePro is now commercialized by Real Time Visualization, a MERL spin-off company.

# Mechanical Design

A MentorBot is an expressive robotic creature with perceptual and motor modalities tailored to natural human communication channels. To facilitate a natural infant–caretaker interaction, the robot is equipped with visual, auditory, and proprioceptive sensory inputs. The motor outputs include vocalizations, facial expressions, and motor capabilities to adjust the gaze direction of the eyes and the orientation of the head. Note that these motor systems serve to steer the visual and auditory sensors to the source of the stimulus and can also display communicative cues.

## Appearance

Appearance is a major factor in MentorBot development. Humans associate better with someone or something with an appealing look. Robots that work with and for people must be designed for a social world. Researchers who seek to promote the positive interaction of people and robots must be especially attentive to the robot's appearance.

### Face and Head

A face is a very complicated structure. It is of such importance to human well-being that a lot of attention has been given to face studies. Most communication among living beings involves the face. The face houses the majority of our sensory organs. The eyes, nose, mouth and ears are not only involved in their primary tasks, but also are involved in the act of communication. The face is a very important part of any believable model. To make a model believable and realistic, a face must project emotions and act as an interface to the vocal communication.

The android head has cameras behind its eyes that follow movements; sophisticated software drives tiny motors under the polymer skin to mimic facial expressions. The MentorBot face will smile, sneer, frown, and even squint. Its twenty-four mechanical muscles react in under one second to produce the copycat visage.

"This is the face for social robotics," said David Hanson, who is building the machine as part of his PhD studies at the Universiity of Texas in Dallas. "The human face is the most natural paradigm for human–computer interactions. This is how we will interact with the computers of tomorrow."

By constructing an emotion exhibiting robot head (Figure 3.8), a humanoid project at Waseda University is taking steps toward human-friendly robots that can emote. The project investigates the confluence of physical and psychological dimensions of intelligence by connecting four modes of sensation to differential equations that continuously compute the emotional state of the robot. The latest version of the robot is called WE-3RIII (Waseda Eye-No. 3 Refined version III) and it has four sensations, including vision, hearing, cutaneous, and smell. To express emotion, the robot is equipped with eyebrows, eyelids, lips, jaw and facial skin that can change colors to express emotions such as anger or embarrassment. It also has cutaneous sensation that allows it to perceive when it is being pushed, stroked, and hit. The robot can also sense warmth near its face and can perceive strong smells such as alcohol, smoke, and ammonia.

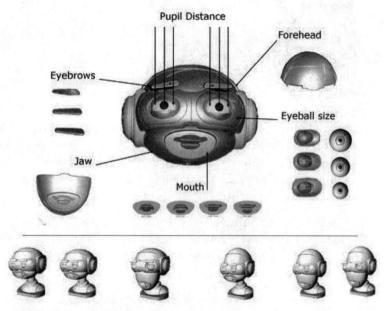

■ **Figure 3.8**  *Humanoid head and face.*

## Neck and Torso

Humanoid robots are highly complex structures that possess degrees of freedom far exceeding the number of inputs from a joystick or a game pad (Figure 3.9).

A human has only one locus of attention: We can only consciously attend to one object or idea at a time. Despite having a large number of joints in our bodies, we carry out a specific task with our locus of attention focusing only on some specific points of our body. Depending on the desired task, the points on which we execute motor command differ.

87

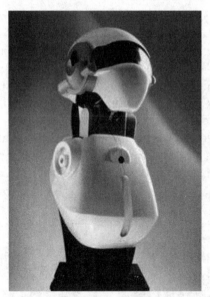

■ **Figure 3.9** *Neck and torso.*

For example, during a task to reach a bottle on a table, our locus of attention is on our hand to accomplish the task. When we lean down, our locus of attention is on our torso. When we kick a ball, our locus of attention shifts to the leg. Human motions are generated by a mixture of conscious and subconscious motion generations.

Conscious motion generations are triggered and supervised by conscious executions of motor command on specific points of the body, which are the loci of attention of the motion, to accomplish the desired force or position of the targeted points.

Subconscious motion generations provide automatic, reactive motions to ensure the stability and safety of the body, such as leg movements for maintaining body balance and reactive motions to prevent injury, which are generated without the supervision of the conscious mind.

### Arm and Shoulder

A small mobile robot equipped with a stereoscopic machine vision system and two manipulator arms that have limited degrees of freedom has been given the ability to perform moderately dexterous manipulation autonomously, under control of an onboard computer (Figure 3.10). The approach taken has been to formulate vision-based control software to utilize the mobility of the vehicle to compensate for the limited dexterity of the manipulator arms. Although the goal was selected visually, it is tracked onboard using information about its shape; in par-

■ **Figure 3.10** *Arm and shoulder.*

ticular, the target is assumed to be a local elevation maximum (i.e., the highest point within a small patch of area).

The mobile robot in question is Rocky 7—a prototype "rover"-type vehicle used in research on robotic-vehicle concepts for the exploration of Mars. The mobility system of Rocky 7 is based on a six-wheel-drive rocker-bogey mechanism that includes two steerable front wheels and four nonsteerable back wheels. One of the manipulator arms is equipped with two independently actuated scoops for acquiring samples. Not counting the motions of the scoops, this arm has two degrees of freedom—shoulder roll and a shoulder pitch. The other manipulator arm is a mast on which is mounted a stereoscopic pair of video cameras that can, if desired, be tipped with a scientific instrument. The mast has three degrees of freedom (shoulder pitch, shoulder roll, and elbow pitch). The mast can be used to position and orient its cameras or place its tip instrument on a target object to acquire a sample or take a reading. Additional stereoscopic pairs of cameras are located at the front and rear ends of the main body of the vehicle.

At the beginning of an operation, one or more target objects a small distance away are selected, and then the robot is commanded to perform autonomously some manipulations that involve the objects. Following the basic approach of using mobility to augment dexterity, the navigation and mobility control subsystems of the vehicle cause the vehicle to maneuver into a position and orientation so that the target lies within the range of one of the manipulators, and then the manipulator control subsystem causes the manipulator to perform the remaining fine positioning and manipulation.

The key to ensuring that the rover reaches its target is for it to move in small steps and lock onto the target by automatically tracking its shape. The computer processes the image data from the stereoscopic camera pairs into an elevation map of the nearby terrain and locates the target on the elevation map. The computer plans the route of the vehicle across the terrain toward the target, using an approximate kinematical model (assuming flat terrain and no slippage of wheels). At frequent intervals along the route, updated elevation maps are generated from newly acquired stereoscopic-image data, the target is identified on the updated maps, and the planned route is corrected accordingly. Here the scale-invariant features (i.e., shape, elevation, and centroids) are tracked: The target is tracked even as its image grows dramatically in size during the final approach, a situation that often causes traditional visual servoing techniques to fail. This process of iterative, vision-based refinement of the route continues until the vehicle arrives at the desired location near the target.

Once the vehicle is in the desired position and orientation relative to the target, the designated manipulator arm is lowered toward the target; tactile sensing is used to signal contact with the target or with the ground adjacent to the target. The manipulator arm is then command-ed to perform the assigned manipulation. Manipulations that Rocky 7 has performed in demonstrations include grasping several small rocks that were initially at a distance of >1 m and placing an instrument on a boulder that was initially at a distance of >5 m.

## Locomotion

Most MentorBots will be mobile. There are three primary means of mobilizing a robot: wheeled, four-legged, and bipedal (two-legged).

### Wheeled Robot

The choice of electric actuators imposes great restrictions in terms of available power and torque (Figure 3.11).

Human muscle has a maximum power density of about 100 W/kg, which is readily achievable with today's "fractional horsepower" electric motors. These can attain peak power/mass ratios of up to 1 kW/kg. Nevertheless, performance comparable to that of muscles remains elusive, because muscles are "direct drive" actuators—they can develop their maximum force and power at the low speeds needed to drive the biological joints directly. Small motors, with a mass of about 1 kg, on the other hand, are severely torque limited, and the peak power is achieved only at high speeds (5000–10,000 RPM). Large direct drive motors, whose intricate geometry cannot be scaled down to small motors, cannot at present reach peak torque/mass ratios of 6 Nm/kg. This calls for a speed reducer (gear) to match the high torque–low velocity requirements of biological or robotic joint motion to the available low torque–high velocity of small motors.

■ **Figure 3.11** *Wheeled robots.*

This has two main consequences: First, virtually all large gear-reducing mechanisms introduce additional dynamics that cannot be ignored. A considerable fraction of the motor power can be consumed by merely accelerating gear inertia. Second, most high-gear ratio mechanisms introduce significant loss of driving torque, up to 61 percent for a high quality three-stage planetary gear head for a gear ratio range of 80:1 to 200:1. This precludes the use of fractional horsepower motors as direct drive force-controlled actuators in legged robots, in an analogous fashion to muscles. In addition, the rotary motion of the motor must be converted into linear motion suitable for a prismatic leg. A ball screw is the only practical transmission that provides both revolute-to-linear motion translation and high efficiency (>95 percent).

Preliminary simulations emphasize that the main design challenge is to maximize the energy added during the short stance phase. In this case, the complete motor–ball screw–spring system must be optimized to this effect at some nominal operating point. Unfortunately, many practical considerations stand in the way of formulating and solving a standard optimization problem. For example, lowering the spring constant can increase the stance time. However, this increases the actuator travel during stance. This in turn requires a longer ball screw (only available at discrete lengths), which increases its moment of inertia (subject to catalog tables), which actually limits the energy added. In addition only a finite number of choices are for suitable motors and ball screw lengths

**91**

and leads. This results in a highly nonlinear and complex optimization problem. To simplify this task, designers discredited the number of available spring constants, and performed an exhaustive (computer) search: The components to maximize the added stance energy (approx. 10 J) were determined as an 80W brushed DC motor (Maxon, 1.3 kg), a 5mm/rev ball screw (RHP), and a 4kN/m spring (custom). The component selection is far from intuitive since the robot–actuator is a fourth-order dynamic system. For example, a higher-power (100W), lower-mass (0.3 kg) brushless motor fared far worse due to the smaller torque at low speed; the velocities where high power is available are never attained during the 0:1s stance time.

### Four-Legged Robot

Dynamically stable legged robots (Figure 3.12) promise to be the mobile platform of choice compared to wheeled systems when it comes to mobility, versatility, and speed, in all but those special cases where a continuous and smooth path of support is provided. However, before legged robots become practical, strong stability properties and autonomous operation are essential. To achieve these goals, two key issues must be considered.

■ **Figure 3.12**   *Four-legged robot.*

First, electrical actuation, instead of hydraulics, must be used as a clean, safe, and cheap technology, suitable for indoor use and autonomous robots. It is clear that replacing the hydraulic actuators on existing robots with electric motors is not a straightforward engineering issue. Since the electric motor's torque/mass ratio is far from that of hydraulic actuators, robot design and control will be fundamentally affected and pose challenging research problems. For example, the actuation strategy of applying a step length change at maximum leg compression for maintaining a constant vertical body oscillation might be a reasonable strategy for powerful hydraulic actuators, but it should be clear that such an approach is no longer advisable with electric actuators. Researchers have verified in simulations and experiments that stable operation cannot be achieved in this manner and propose a new stabilizing continuous controller with an average power consumption of only 80W.

Second, in model-based control using verified dynamic models of robot, dynamic interaction with the ground and the actuators themselves is mandatory if predictable performance is desired. This is especially crucial with electric actuator models, which impose considerable operational limits. In addition, currently a gulf exists between experimental and theoretical research in this field. Due to the lack of supporting experimental work and insufficient modeling efforts, much of the recent analytical literature reviewed operates in a vacuum; it is not clear which model simplifications and actuator models are valid, and which analytical methods and results are meaningful in practice. The benefits of empirically relevant analytical work are considerable and could lead to new and better control laws, a formal procedure for gain selection to improve stability, a better understanding of the complicated dynamics of legged robots in general, and in the future, better guidelines for robot design.

**Bipedal Robot**

The final goal is the realization of human-type autonomous robot (Figure 3.13). At first, a new 3-D precise simulation model was developed that modeled the collision between the foot and floor. Through an investigation of the various control algorithms by the simulator, the importance of environmental uncertainty was discovered. Therefore, a bipedal walking system based on force control was proposed. This approach seems more suitable for the legged locomotion system than for a trajectory-controlled one.

The hardware and software control architectures were designed to meet the challenge of the real-time processing of visual signals (approaching 30 Hz) and auditory signals (8 kHz sample rate and frame windows of 10 ms) with minimal latencies (less than 500 ms). The high-level perception system, the motivation system, the behavior system, the motor skill system, and the face motor system execute on four Motorola 68332 microprocessors running L, a multithreaded LISP developed in the lab.

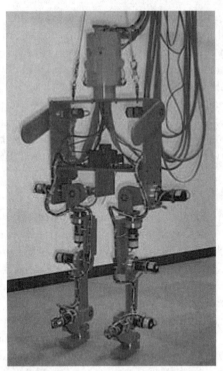

■ **Figure 3.13** *Bipedal mechanism.*

### The Motor System

The motor system arbitrates the robot's motor skills and expressions. It consists of four subsystems: the *motor skills system*, the *facial animation system*, the *expressive vocalization system*, and the *oculomotor system*. Given that a particular goal and behavioral strategy have been selected, the motor system determines how to move the robot so as to carry out that course of action. Overall, the motor skills system coordinates body posture, gaze direction, vocalizations, and facial expressions to address issues of blending and sequencing the action primitives from the specialized motor systems.

Current designs use small sixteen-channel motion control modules. The motor drivers are standard FET H-bridges; recent advances in FET process technology permit surprisingly low RDS on losses, and switching at relatively low (1–10kHz) frequencies reduces switching losses.

Therefore, the power silicon (and thus the package as a whole) can be reduced in size. Audible hum and interference (which is completely unacceptable in an organic looking robot) is normally due to the low switching frequency and can be eliminated by using a variable-mean spread-spectrum control signal, rather than traditional PWM. Each channel supports current feedback, encoder feedback, and analog feedback, and the system is controlled by a custom SoC motion controller with an embedded soft processor core implemented in FPGA.

# Navigation

Autonomous robots need the ability to move purposefully without human intervention, so they need navigation. Navigation from point A to point B while avoiding any obstacles in the path can be accomplished with odometry, sonar, vision, or a combination of all three.

## Odometry (Dead-Reckoning) Navigation

Odometry provides good short-term accuracy, is inexpensive, and allows very high sampling rates. However, the fundamental idea of odometry is the integration of incremental motion information over time, which leads inevitably to the accumulation of errors. In particular, the accumulation of orientation errors will cause large position errors, which increase proportionally with the distance traveled by the robot. Nevertheless, it is widely accepted that odometry is a very important part of a robot navigation system and that navigation tasks will be simplified if odometry accuracy can be improved:

■ Odometry data can be fused with absolute position measurements (using other techniques) to provide better and more reliable position estimation.

■ Odometry can be used between absolute position updates with landmarks. Given a required positioning accuracy, increased accuracy in odometry allows for less frequent absolute position updates. As a result, fewer landmarks are needed for a given travel distance.

■ In some cases, odometry is the only navigation information available, as when there are insufficient landmarks in the environment or when another sensor subsystem fails to provide usable data.

### Sensors for Dead Reckoning

Dead reckoning (derived from "deduced reckoning" from sailing) is a simple mathematical procedure for determining the present location of a vehicle by advancing some previous position through known course and velocity information over a given length of time.

At present, the vast majority of land-based mobile robots rely on dead reckoning to form the backbone of their navigation strategy. They use other navigation aids to eliminate accumulated errors.

## Odometric Sensors

The simplest form of dead reckoning is often termed as odometry. This implies that the vehicle displacement along the path of travel is directly derived from some on-board odometer. A common means of odometric measurement involves optical encoders directly coupled to wheel axles.

Since a large majority of mobile robots rely on motion by means of wheels or tracks, a basic understanding of sensors that accurately quantify angular position and velocity is an important prerequisite for dead reckoning using odometry.

Some common rotational displacement and velocity sensors in use today are:

- Brush encoders
- Potentiometers
- Optical encoders
- Magnetic encoders
- Inductive encoders
- Capacitive encoders

The most popular type of rotary encoder for mobile robots is the *incremental* or *absolute* optical encoder.

## Optical Encoders

At the heart of an optical encoder (Figure 3.14) lies a miniature version of the *break-beam proximity sensor*. Here, a focused beam of light aimed at a matched photo detector (e.g., a slotted optic switch) is periodically interrupted by a coded opaque and transparent pattern on a rotating intermediate disk attached to the shaft of interest. The advantage of this encoding scheme is that the output is inherently digital, which results in a low-cost reliable package with good noise immunity.

The incremental optical encoder measures rotational velocity and can infer relative position. The absolute model, on the other hand, measures angular position directly and can infer velocity. If nonvolatile position information is not a requirement, incremental encoders are usually chosen on grounds of lower cost and simpler interfacing compared to absolute encoders.

■ **Figure 3.14**   *Optical encoder.*

Discrete elements in a photovoltaic array are individually aligned in break-beam fashion with concentric encoder tracks, creating in effect, a noncontact implementation of a commutating brush encoder. Having a dedicated track for each bit of resolution results in a larger disk (relative to incremental designs), with a corresponding decrease in shock and vibration tolerance. Very roughly, each additional encoder track doubles the resolution and quadruples the cost.

### Incremental Optical Encoders

The single-channel *tachometer encoder* is the simplest type of incremental encoder. It is basically a mechanical light chopper that produces a certain number of pulses per shaft revolution. Increasing the number of pulses per revolution increases the resolution (and cost) of the encoder. These devices are especially suited as velocity feedback sensors in medium- to high-speed control systems. However, they run into noise and stability problems at very slow velocities due to quantization errors. In addition to these instabilities, the single-channel tachometer encoder is incapable of detecting the direction of rotation and thus cannot be used as a position sensor.

To overcome these problems, a slightly improved version of the encoder called the *phase-quadrature incremental encoder* is used. Here, a second channel, displaced from the first, is added. This results in a second pulse train that is 90 degrees out of phase with the first pulse train. Now, decoding electronics can determine which channel is leading the other and hence determine the direction of rotation, with the added benefit of increased resolution.

Since the output signal of these encoders is incremental in nature, any resolution of angular position can only be relative to some specific reference, as opposed to absolute. For applications involving continuous 360-degree rotation, such a reference is provided by a third channel as a special index output that goes high once for each revolution of the shaft. Intermediate positions are then specified as a displacement from the index position. For applications with limited rotation, such as the back-and-forth motion of a pan axis, electrical limit switches can be used to establish a home reference position. Repeatability of this homing action is often broken into steps. The axis is rotated at reduced speed in the appropriate direction until the stop mechanism is encountered. Rotation is then reversed for a short predefined interval, after which the axis is then slowly rotated back to the stop position from this known start point. This usually eliminates inertial loading that could influence the final homing position (This two-step approach can be observed in the power-on initialization of stepper-motor positioners in dot-matrix printer heads.)

Interfacing an incremental encoder to a computer is not a trivial task. A simple state-based interface is inaccurate if the encoder changes direction at certain positions, and false pulses can result from the interpretation of the sequence of state changes.

A popular and versatile encoder interface is the HCTL 1100 motion controller chip made by Hewlett Packard. It performs accurate quadrature decoding of the incremental wheel encoder output and provides important additional functions such as closed-loop position control, closed-loop velocity control, and 24-bit position monitoring.

At a cost of $40, it is a good candidate for giving mobile robots navigation and positioning capabilities.

### Absolute Optical Encoders

These are typically used for slower rotational applications that do not tolerate loss of positional information when there is a temporary power interruption. They are best suited for slow and/or infrequent rotations such as steering angle encoding, as opposed to measuring the high-speed continuous rotation required for calculating displacement along the path of travel.

Instead of the serial bit streams of incremental designs, absolute encoders provide a parallel word output with a unique code pattern for each quantized shaft position. The most common coding scheme is the Gray code. This code is characterized by the fact that only one bit changes at a time, thus eliminating (a majority of the) asynchronous ambiguities caused by electronic and mechanical component tolerances.

A potential disadvantage of absolute encoders is their parallel data output, which requires more complex interface due to the large number of electrical leads.

### Systematic/Nonsystematic Odometer Errors

Odometry is based on the assumption that wheel revolutions can be translated into linear displacement relative to the surface. This assumption is of limited validity since displacement errors can easily be caused by wheel slippage. Error sources are grouped into two distinct categories, systematic and nonsystematic.

- Systematic Errors
  - Unequal wheel diameters
  - Actual wheelbase differs from nominal wheelbase
  - Misalignment of wheels
  - Finite encoder resolution
  - Finite encoder sampling rate
- Nonsystematic Errors
  - Travel over uneven floors
  - Travel over unexpected objects on the floor
  - Wheel slippage due to slippery floors
  - Over acceleration
  - Fast turning (skidding)
  - External forces (interaction with external bodies)
  - Internal forces (castor wheels)

Making a clear distinction between these two error categories is of great importance for the effective reduction of odometry errors. On most smooth indoor surfaces, systematic errors contribute much more to odometry errors than do nonsystematic errors. However, on rough surfaces with significant irregularities, nonsystematic errors are dominant. The problem with nonsystematic errors is that they appear unexpectedly and can cause large position errors.

### Measurement of Systematic Errors

One widely used but fundamentally unsuitable method to measure systematic errors is the unidirectional square-path test (USPT). For this test, the robot must be programmed to traverse the four legs of a square path. The path will return the vehicle to the starting area but, because of odometry and controller errors, not precisely to the starting position. The systematic error is then calculated from the end position error. An improved version of this is a bidirectional USPT that involves the robot traveling around a square path in both clockwise and counterclockwise directions.

**Measurement of Nonsystematic Errors**

Due to the unpredictable nature of nonsystematic errors, it is very difficult (perhaps impossible) to design a generally applicable quantitative test procedure for them. One proposed approach is to compare the susceptibility to nonsystematic errors of different vehicles. This method uses the bidirectional square-path test but in addition introduces artificial bumps. The robot's susceptibility to nonsystematic errors is indicated by the return orientation error and not the return position error.

**Reduction of Odometry Errors**

The accuracy of odometry in mobile robots depends to some degree on their kinematics design and on certain critical dimensions. Some design-specific considerations that affect dead-reckoning accuracy are:

- Vehicles with a small wheelbase are more prone to orientation errors than vehicles with a larger wheelbase. For example, the differential drive LabMate robot from TRC has a relatively small wheelbase of 340 mm. As a result, dead reckoning is limited to a range of about 10m before a new position "reset" is necessary.
- Vehicles with caster wheels that bear a significant portion of the overall weight are likely to induce slippage when reversing direction (the "shopping trolley effect").
- Wheels used for odometry should be "knife-edge" thin and not compressible. The ideal wheel is made of aluminum with a thin layer of rubber for better traction. In practice, this is only practical for lightweight robots.

Another general observation is that errors associated with wheel slippage can be reduced by limiting the vehicle's speed during turning, and by limiting accelerations.

**Reduction of Systematic Odometry Errors**

It is generally possible to improve odometry accuracy by adding a pair of "knife-edge," non-load-bearing *encoder wheels*, as shown in Figure

3.15. Since these wheels are not used for transmitting power, they can be small and lightweight.

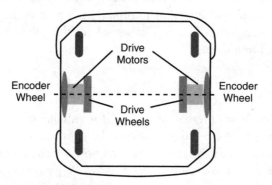

■ **Figure 3.15**   *Conceptual drawing of a set of encoder wheels for a differential drive vehicle.*

An alternative approach is the use of an *encoder trailer* with two encoder wheels. This approach is often used for tracked vehicles, where it is virtually impossible to use odometry, because of the large amount of slippage between the tracks and the floor during turning.

A digital compass can be useful for mobile robot navigation, especially for a small robot such as the PPRK, which lacks wheel encoders and hence built-in odometry and dead reckoning.

Another approach to improving odometric accuracy without any additional devices or sensors is based on careful calibration of the mobile robot. Systematic errors are inherent properties of each individual robot and they change very slowly as a result of wear or of different load distributions. This technique of reducing errors requires high precision and accurate calibration, since minute deviations in the geometry of the vehicle or its part may cause substantial odometric errors. As a result, this technique is time consuming.

### Reduction of Nonsystematic Odometry Errors

*Mutual referencing* is a method whereby two robots measure their positions mutually. When one robot moves to another place, the other remains still, observes the motion, and determines the first robot's new position. In other words, one robot localizes itself with reference to a fixed object (the stationary robot). However, this limits the efficiency of the robots.

A unique way for reducing odometry errors even further is *internal position error correction*. Here, two mobile robots mutually correct their odometry errors on a continuous basis. To implement this method, both robots must measure their relative distance and bearing continuously. Conventional dead reckoning is used by each robot to determine its heading and direction information, which is then compared to the heading and direction observed by the other robot to reduce errors to a minimum.

## Inertial Navigation

Inertial navigation is an alternative method for enhancing dead reckoning. The principle of operation involves the continuous sensing of minute accelerations in each of the three directional axes and integrating these over time to derive velocity and position. A gyroscopically stabilized sensor platform is used to maintain consistent orientation of the three accelerometers throughout this process.

Although this method is simple in concept, the specifics of implementation are rather demanding, mainly because of error sources that affect the stability of the gyros used to ensure correct attitude. The resulting high manufacturing and maintenance costs of this method have usually made it impractical for mobile robot applications. For example, a high-quality *inertial navigation system* (INS), as would be found in a commercial airliner, will have a typical drift of about 1850 m per hour of operation. High-end INS packages used in ground applications have shown performance of better than 0.1 percent of distance traveled. However, since the development of laser and optical fiber gyroscopes, INS is becoming more suitable for mobile robot applications.

One advantage of inertial navigation is its ability to provide fast, low-latency dynamic measurements. Also, INS sensors are self-contained, nonradiating, and nonjammable. The main disadvantage however, is that the angular rate data and the linear velocity rate data must be integrated once and twice (respectively), to provide orientation and linear position, respectively.

## Accelerometers and Gyroscopes

*Accelerometers* have a very poor signal-to-noise ratio at lower accelerations (i.e., during low-speed turns). They also suffer from extensive drift, and they are sensitive to uneven ground, because any disturbance from a perfectly horizontal position will cause the sensor to detect the gravitational acceleration $g$. Even tilt-compensated systems indicate a position drift rate of 1 to 8 cm/s, depending on the frequency of acceleration changes. This is an unacceptable error rate for most mobile robot applications.

*Gyroscopes* provide fundamental angular rate and accelerometers provide velocity rate information. Dynamic information is provided through direct measurements. The main problem with gyroscopes is that they are usually very expensive (if they are to be useful for navigation), and they need to be mounted on a very stable platform.

## Physical Scales

The physical scale of a device's navigation requirements can be measured by the accuracy to which the mobile robot needs to navigate—this is the *resolution of navigation*. These requirements vary greatly with application, however a first order approximation of the accuracy required can by taken from the dimensions of the vehicle itself. Any autonomous device must be able to determine its position to a resolution within at least its own dimensions, in order to be able to navigate and interact with its environment correctly.

At the small end of the scale are robots just a few centimeters in size, which will require high-precision navigation over a small range (due to energy supply constraints), while operating in a relatively tame environment. At the other end of the scale are jumbo jet aircraft and ocean-going liners, each with some sort of autopilot navigation that requires accuracy to a number of meters (or tens of meters), over a huge (i.e., global) range, in somewhat more rugged conditions.

To help categorize this scale of requirements, we use three terms:

- **Global navigation.** The ability to determine one's position in absolute or map-referenced terms, and to move to a desired destination point.
- **Local navigation.** The ability to determine one's position relative to objects (stationary or moving) in the environment, and to interact with them correctly.
- **Personal navigation.** Being aware of the positioning of the various parts that make up oneself, in relation to each other and in handling objects.

In a jet autopilot, global navigation is the major requirement for cruising between continents. Local navigation becomes necessary when the aircraft is expected to fly in crowded airways, or on approach to the runway on landing. Personal navigation is not an issue, as the vehicle is, fundamentally, one object, and should never come into contact with any other significant objects while under autonomous control.

The "micro" robot, on the other hand, is almost exclusively interested in personal and local navigation. Such devices are rarely concerned with their position globally, on any traditional geographic scale. Instead, their requirements are far more task based—they are concerned with their immediate environment, relative to any objects relevant in the successful completion of their task. This involves personal navigation, when the robot is in contact with other objects, and local navigation for actual movement.

## Sonar Navigation

Ultrasonic sensors measure the distance or presence of a target object by sensing a reflected sound wave, above the range of hearing, from an object in the robot's field of view, and then measuring the time for the sound echo to return. Knowing the speed of sound, the sensor circuitry can determine the distance of the object from the transducer element. Bats use this approach to fly with unerring accuracy, sense tiny flying insects, and feed on them in flight.

Like bats, the transducer transmits at a frequency much higher than that of human hearing. Some people may hear a clicking as the transducer is energized, but not even dogs can hear this ultrasonic sound of 50,000 Hertz. The measurement sampling period is 50 m Sec for a 20 Hz update, thus allowing rapid distance measuring data for the PCR.

Most high-end, research autonomous mobile robots have no more than seven or nine transducers in fixed positions, covering the same 180 degrees from hard left to hard right. Having more unique measurement angles is better due to the natural reflective behavior of sonar. A fixed, high-count sonar array approach is expensive and complicated, because multiple sonar tend to interfere with each other and hence reduce mutual effectiveness. Even with multiplexing, hardware and software management of their interaction is generally costly. Additionally, only a limited number of sonar transducers can physically fit on the circumference of a robot. Typically seven to nine is common; more sensors require a larger diameter robot. Robots with an outside diameter of over 20 inches are generally unsuitable for home or office use.

The problem of map building is even harder than that of map-based position estimation, where, in addition to estimating the position of a robot, the map has to be estimated simultaneously. Based on previous experiences in map building and global position estimation, researchers developed an approach to concurrent map building and localization. The approach uses the EM-algorithm to estimate the most likely map given the robot's observations.

The approach is extremely cost effective, can build a map of the world, can react to stimuli from many directions, and has high resolution. This robust sweeping sonar system uses many discrete angles (twenty-five), generally two to three times as many as autonomous mobile robots costing much more. High-count arrays, while giving higher resolution, are more expensive to effectively implement.

The rotational displacement sensors derive navigation data directly from wheel rotation, thus they are inherently subject to problems arising from wheel slippage, tread wear, or improper tire inflation. Doppler and inertial navigation techniques are sometimes employed to reduce the effects of such error sources.

### Doppler Sensors

The principle of operation in Doppler sensors is based on the Doppler shift in frequency observed when radiated energy reflects off a surface that is moving with respect to the emitter.

Most implementations used for robots employ a single forward-looking transducer to measure ground speed in the direction of travel. An example of this is taken from the agricultural industry, where wheel slippage in soft, freshly plowed fields can seriously interfere with the need to release seed at a rate proportional to vehicle advance.

A typical implementation uses a microwave radar sensor, which is aimed downward (usually 45 degrees) to sense ground movement as shown in Figure 3.16.

$$V_A = \frac{V_D}{\cos\alpha} = \frac{cF_D}{2F_0 \cos\alpha}$$

where
$V_A$ = actual ground velocity along path
$V_D$ = measured Doppler velocity
$\alpha$ = angle of declination
$c$ = speed of light
$F_D$ = observed Doppler shift frequency
$F_0$ = transmitted frequency

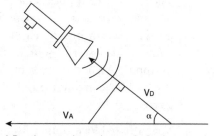

A Doppler ground-speed sensor inclined at an angle $\alpha$ as shown measures the volocity of component $V_0$ of true ground speed $V_A$. (Adapted from Schultz, 1993)

■ **Figure 3.16**   *Radar sensor.*

Errors in detecting true ground speed can arise from vertical velocity components introduced by vehicle reaction to the ground surface and uncertainties in the angle of incidence. An interesting scenario resulting in erroneous operation would involve a stationary vehicle parked over a stream of water.

### Navigation Reference

Following on, and going in hand with, the scale of navigation requirements, is the frame to which relative position fixing is performed. The two terms used here are, understandably, *absolute* and *relative*. What isn't so obvious, however, is what relative is relative to, and where absolute is absolute from, because these terms are somewhat context sensitive.

In terms of position fixing, absolute implies finding one's position relative to an absolute origin; a fixed stationary point common to all position fixes across the range of navigation. Hence, in global navigation, there should be one such point on the planet to which all fixes are relative. In local navigation, the absolute origin is some fixed point in the robot's environment, and in personal navigation the origin can be viewed as the center of the robot itself.

A relative position fix when navigating globally, taken relative to some other reference point (environment-relative), is analogous to the absolute position fix in local navigation. Likewise, a position fix taken relative to the same robot's own position at some other point in time (self-relative), is like the personal absolute position fix. Through knowledge of the absolute reference frame (typically using a map), absolute position fixes in one navigation domain can be transformed into position fixes in another.

## Vision-Based Navigation

The real-world applications envisaged in most current research projects demand very detailed sensor information to provide the robot with good environment-interaction capabilities. Visual sensors are potentially the most powerful source of information among all the sensors used on robots. Hence, at present, it seems that high-resolution optical sensors hold the greatest promise for mobile robot positioning and navigation.

Vision-based localization using landmarks complements robot odometry. Stereo data from binocular CCD video cameras is used to build an occupancy grid map of the environment for navigation. Obstacle detection is based on stereo vision data using sonar as well as infrared sensors. Localization and map building is a key component of a mobile robot system. 3-D landmarks based on scale-invariant image features are tracked over frames to correct the odometry errors to improve robot ego-motion. A 3-D map of the landmarks is obtained and is used for robot localization.

The results strongly indicate that vision-based navigation possesses a corrective or feedback trait that produces an upper bound on the movements to the goal position. In contrast to the odometry-based movement, the inability to produce the required turn as requested leads to the "drift" between successive navigation movements. This causes the displacement from the goal position to drift and increase without bound.

Vision-based positioning or localization uses the same basic principles of landmark-based and map-based positioning but relies on optical sensors rather than ultrasound, dead reckoning, and inertial sensors. The environment is perceived in the form of geometric information such as landmarks, object models, and maps in two or three dimensions. Localization then depends on the following inter-related considerations:

- A vision sensor (or multiple vision sensors) should capture image features or regions that match the landmarks or maps.
- Landmarks, object models, and maps should provide necessary spatial information that is easily sensed.

The primary techniques employed in vision-based positioning are:

- Landmark-based positioning
- Model-based approaches
- Three-dimensional geometric model-based positioning
- Digital elevation map-based positioning
- Feature-based visual map building

Although it seems a good idea to combine vision-based techniques with methods using dead reckoning, inertial sensors, and ultrasonic and laser-based sensors, applications under realistic conditions are still scarce.

The most common optical sensors include laser-based range finders and photometric cameras using CCD arrays. However, due to the volume of information they provide, the extraction of visual features for positioning is far from straightforward. Many techniques have been suggested for localization using vision information, the main components of which are:

- Representations of the environment
- Sensing models
- Localization algorithms

Clearly, vision-based positioning is directly related to most computer vision methods, especially object recognition. So, as research in this area progresses, the results can be applied to vision-based positioning.

## Power Source

One of the single most expensive components that goes into the make-up of an amateur mobile robot is the power source. This means, of course, the batteries that supply the current necessary to power on-board motors, sensors, and processors unless someone has worked the bugs out of cold fusion. Table 3.1 compares various battery sources.

**Table 3.1**  *Comparing Rechargable Technologies*

| Characteristic | Types | | | | |
|---|---|---|---|---|---|
|  | Sealed Lead-Acid | Nickel Cadmium* | Nickel Metal Hydride* | Lithium Ion* | Lithium Metal* |
| Avg Operating Voltage (V) | 2 | 1.2 | 1.25 | 3.6 | 3.0 |
| Energy Density (Wh/kg) | 35 | 45 | 55 | 100 | 140 |
| Vol. Efficiency (Wh/L) | 85 | 150 | 180 | 225 | 300 |
| Cost ($/Wh) | 0.25 to 0.50 | 0.75 to 1.5 | 1.5 to 3.0 | 2.5 to 3.5 | 1.4 to 3.0 |
| Memory Effect ? | No | Yes | No | No | No |
| Self-Discharge Rate (% month ) | 5 to 10 | 25 | 20 to 25 | 8 | 1 to 2 |
| Temperature Range (°C) | 0 to +50 | –10 to +50 | –10 to +50 | –10 to +50 | –30 to +55 |
| Environmental Concerns ? | Yes | Yes | No | No | No |

* Based on AA-size Cell

## Battery Choices

In general, designers and end users want battery packs with the lowest possible weight and highest possible capacity crammed into the smallest possible volume at the lowest possible cost. These characteristics are especially important as the trend to miniaturization in portable products accelerates. While no single rechargeable battery type can optimize all of these factors, there are some obvious trade-offs among available technologies.

Rechargeable batteries for MentorBots come in five flavors: sealed lead-acid, nickel cadmium, nickel metal hydride, lithium ion, and lithium metal.

At the low end of the cost spectrum is the sealed lead-acid (SLA) battery. This chemistry is mature and reliable, but it's also at the low end of the scale for volumetric efficiency (energy per unit volume) and energy density (energy per unit weight). You can expect an average of 85 Wh/L and 35 Wh/kg, respectively. Sealed lead-acid batteries were used in early hand-held systems but aren't widely used in smaller robots. Most lead-acid batteries cannot be rapidly recharged (in, say, less than three hours) because of possible thermal damage, although special rapid-charge lead-acid types are available. Lead, of course, is a well-known pollutant and lead-acid batteries must be recycled.

Still near the lower end of the cost range but significantly better than lead-acid in volumetric efficiency and energy density (120 Wh/L and 36 Wh/kg) is the "standard" sintered-metal-electrode nickel cadmium (NiCd) cell. With the introduction a few years ago of sponge-metal electrode technology, the volumetric efficiency of NiCd cells jumped to 150 Wh/L and energy density rose to 50 Wh/kg.

An important parameter for rechargeable batteries is their self-discharge rate, the rate at which the battery loses charge while not in use. Self-discharge rate can become an issue for end-users who use a particular portable system infrequently, yet want to be able to rely on it when necessary. Self-discharge for NiCd batteries is moderate compared to other types.

Nickel cadmium is the most familiar secondary-battery technology and is therefore widely used. Charging circuits are relatively simple and charging is relatively rapid, but care must be taken to avoid extended periods at high temperatures during charging.

Although NiCd cells are popular and well-understood, they have some drawbacks. First, most batteries of this type available today exhibit *memory effect*, which is the loss of capacity that results when the battery is recharged before it is fully discharged. Second, these batteries contain cadmium, a hazardous substance, and must be recycled in most areas.

Another common rechargeable alternative is nickel metal hydride (NiMH) batteries. NiMH cells offer increased volumetric efficiency (190 Wh/L) over even the most advanced NiCd types. Energy density is also better than that of NiCd, at 55 Wh/kg. Open-circuit voltage for NiMH cells is 1.25 V, which is identical to that of NiCd cells. As a result, some designers have used NiMH batteries as drop-in replacements for NiCd packs. However, NiMH cells cost significantly more than NiCd cells (up to twice as much, depending on form factor) and require special charging circuits that are substantially different from the relatively simple ones used for NiCd. For NiMH, charging time, rate, and temperature must be accurately controlled. NiMH batteries also have the highest self-discharge rate of any of the batteries discussed here, making their use in some types of portable systems of questionable value.

Because NiMH cells contain no hazardous substances, disposal is not an issue. They also do not exhibit the memory effect associated with NiCd types, but their longevity is directly related to depth of discharge.

A third rechargeable option is lithium-based battery systems, which have always been considered attractive because of their high level of electrochemical performance. On the other hand, safety and environmental considerations originally required complex and costly construction techniques and safety systems. These factors raised costs and restricted early lithium batteries to critical military applications. Newer lithium-based systems have overcome the safety and environmental obstacles and are, in general, the most efficient types available.

The two lithium-based systems available today are lithium ion and lithium metal. Although both types exhibit the overall advantages of lithium-based systems, they differ in some important respects specifically related to portable applications.

Lithium-ion rechargeable batteries were first used in small video camcorders and are now seeing some use in other portable applications. The relatively high voltage (3.6 V) of the lithium-ion cell offers the advantage of fewer cells being required to achieve a given voltage.

One major drawback to lithium-ion technology is its relatively high cost/performance ratio. The cost per watt-hour of a lithium-ion cell is significantly higher than that of other types, but some performance figures are not in proportion (Table 3.1). For example, the volumetric efficiency of a lithium-ion cell is significantly less than that of a lithium-metal cell and only slightly better than that of a nickel-cadmium cell. Energy density is about 64 percent better than NiMH, but at a potentially greater cost penalty. Energy density is also 36 percent below that of lithium metal.

Another shortcoming of the lithium-ion system is its primarily nonlinear discharge characteristic. Typically, an AA-sized lithium-ion battery, discharged at a rate of 250 mA, will drop in voltage from 4.3 V (fully charged) to about 3 V in about 90 minutes. The voltage will remain at 3 V for the next 90 minutes and then drop off rapidly to 2 V, at which point the battery is considered discharged. Depending on the design of the system, this discharge characteristic can be troublesome in some portable applications that require a minimum voltage for operation.

The more recently developed rechargeable lithium-metal (Li/Li + $MnO^2$) batteries offer energy density and volumetric efficiency unmatched by any other battery type. Lithium metal cells, exemplified by Tadiran's In-Charge AA-size cells, have energy density of 140 Wh/kg and volumetric efficiency of 300 Wh/L. The cells are totally safe and are immune to practically all types of physical or electrical abuse conditions. The increased safety factor is due primarily to a Tadiran-patented fail-safe, self-quenching electrochemical system and a built-in safety vent.

Lithium-metal batteries have no memory effect and have the lowest self-discharge rate of all rechargeables. A typical lithium-metal battery stored at room temperature (70°F, 20°C) retains 85 percent of its capacity after one year. A portable system powered by a charged lithium-metal battery will always be available for use regardless of how long it sits idle. Operating temperature range (–30°C to +55°C) is also greater than for other battery types.

The discharge curve of a lithium-metal cell is practically flat. At a 250-mA discharge rate, after a brief (about 10-minute) drop from the fully charged voltage of 3.4 V, the voltage remains at 2.8 V more than 3 hours, dropping off to 2 V (the "discharged" point) after that. This also means that when two cells are used in series (to create a battery with a nominal voltage of 6 V) the voltage remains above 4 V for the entire discharge cycle. That 4-V level is the minimum operating voltage for many portable systems, including many cellular phones.

Lithium-metal batteries are capable of delivering up to 2 A of current under continuous or pulse demand. The latter is especially important in cellular phones, where in a typical 600-mW unit, current demands can jump from a standby current of 40 mA to a 0.6-ms talk pulse of 1.4 A with a 200-mA floor between pulses. Under these conditions, the mean talking current is 333 mA. A four-cell (AA cells) lithium-metal battery pack with a capacity of 1600 mAh, can provide nearly four hours of talk time combined with over 13 hours of standby time between recharges. This is accomplished in a four-cell battery pack that weighs only 68 g (2.4 oz), which is 70 percent of the weight of an equivalent 6-V NiMH pack and about 60 percent of a lithium-ion pack.

### Nickel-Cadmium (NiCd)

NiCd chemistry should be considered first, because this is probably the most popular battery. When properly charged and discharged, a NiCd battery can maintain its capacity through 500 charge cycles and last more than 5 years. Although NiCds will discharge if left "on the shelf," they can be renewed to near 100 percent of capacity after a few charge cycles. However, if the robot must function in really cold weather, another type of battery should be used. When NiCd batteries are charged at temperatures below 0°C, the gas absorption reaction is not adequate and the battery may be damaged.

### Charging

One advantage of NiCds is that they can be charged very rapidly without damaging them—provided the charge process is carefully controlled to avoid overcharging and the battery pack is well ventilated to avoid overheating. A charge method known as –DV (see the chart in Figure 3.17) is used to charge cells in as little as one hour.

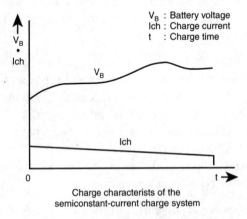

Charge characterists of the
semiconstant-current charge system

■ **Figure 3.17**   *NiCd rapid charge chart.*

The simplest charging method is the "semi-constant current charge." In this scheme, a series resistor is used in the charge path to stabilize the charge current. The battery is charged for about 15 hours. Single-cell battery voltage, $V_C$, is set to 1.45 V, so the battery voltage is determined by a $V_B = V_C \times N$, where $N$ is the number of cells used. Using this method, the battery will be charged at a 0.1 CmA rate. Charge current ($I_{ch}$) is given by:

$$I_{ch} = (V_o - V_b)/R$$

Working through an example, if a six-cell battery pack composed of AA cells that are rated at 1,000 mAH is used, the parameters will be calculated as follows:

$$V_b = 1.45 \times 6 \text{ or } 8.7 \text{ V}$$

This is the fully charged state. It is desirable to charge the battery at a 0.1 C rate. Since robot batteries are generally rated at 1,000 mAH, 100 mA of current will be supplied to the battery during the charge cycle. The resistor should be (8.7 V − 7.2 V)/0.1 A = 15 ohms. The 7.2 V is derived from multiplying the rated cell voltage of 1.2 volts times the number of cells in the battery pack (6).

The idea is that by the time the cells are fully charged, the current will drop to nearly zero. Using this method will help prevent overcharging and overheating of the battery pack, thus preventing damage. Be sure to choose a resistor large enough to handle the power needed. Semi-constant current charging is a safe, inexpensive way to charge NiCads.

### Discharging

A system must be designed to discharge the battery pack at a 0.1 to 0.5 C rate for highest efficiency. Overheating a battery at high discharge rates will lead to deterioration in battery performance. In any case, it's not a good idea to discharge a battery at more than 2 C rates for any extended period. Overdischarging *will* damage the battery pack.

Use this method to determine minimum voltage on the cells.

- Number of arranged serially batteries: 1 to 6
- (Number of batteries × 1.0) V: 7 to 20
- (Number of batteries − 1) × 1.2) V

Make sure a way to monitor battery voltage has been provided in a robot application and that the power is cut when minimum discharge levels are reached.

### Safety

Battery safety consists of a few simple and straightforward rules:

- Don't disassemble the individual cells.
- Never short-circuit these cells.
- Observe polarity when charging and discharging...don't intentionally reverse polarity on these batteries.
- Recycle nickel-cadmium batteries, don't throw them in the landfill. Cadmium is a nasty contaminant.
- Don't attempt to solder directly to the case of the cell. The safety vent or other components could be damaged.

## Sealed Lead-Acid (SLA)

Nominal voltage on SLAs is 1.75 volts per cell under loaded conditions and cells can be charged to as high to 2.45 volts. Probably the most popular charging method for this chemistry is the "constant voltage" method (Figure 3.18).

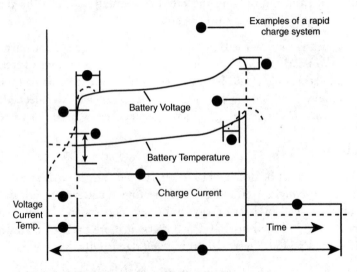

Examples of a rapid charge system

Battery Voltage

Battery Temperature

Charge Current

Voltage
Current
Temp.

Time

■ **Figure 3.18**   *Rapid charging cycle.*

Apply a voltage of 2.45 V/cell and allow the battery to charge to this value. Six cells make up a 12 V battery, so charge voltage would be set to 14.7 V. The power source must be adequate to supply the current the battery demands or the power supply can "crowbar" to protect its circuits using this method. Manufacturers recommend that current be limited to 0.4 C at the beginning of the charge cycle. Charge time will generally be less than 3 hours depending on the state of discharge.

One key feature for maintaining long battery life in an SLA is watching the "depth of discharge" of the battery as it goes about its daily life (Figure 3.19).

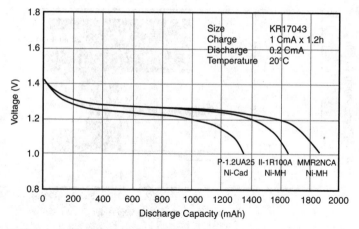

**■ Figure 3.19** *SLA discharging.*

If an application is designed to discharge no more than 30 percent of capacity, the battery can easily be "cycled" a thousand times. Conversely, to wear a battery out fast, deep discharge it (50 percent) often—it will soon be on its way it to the recycler.

It is still worthy of note that, unlike their NiCad and nickel-metal brethren, lead-acid batteries can be deeply discharged and still come back for more. Doing this just reduces the number of charge cycles that a given battery will endure.

As with other battery chemistries, storing lead-acids will allow them to self-discharge over time. This is no problem if they are not left on the shelf too long. A few charge–discharge cycles will bring the battery back up to full capacity. However, if the battery remains on the shelf uncharged for too long, sulfation of the plates can occur, leading to reduced capacity. Batteries in storage should be recharged at least annually, depending on storage temperature.

The state of discharge can be determined by measuring the open circuit voltage of the battery. Figure 3.20 shows the residual capacity vs. the open circuit voltage of the battery.

All the safety points raised in the section on NiCad batteries should be observed here. Additionally, while sealed lead-acid cells don't release a lot of hydrogen gas when charging, they do release small amounts depending on temperature and charge rates. Therefore, do not use this battery in tightly enclosed applications where gases can collect. Recycle dead lead-acid batteries.

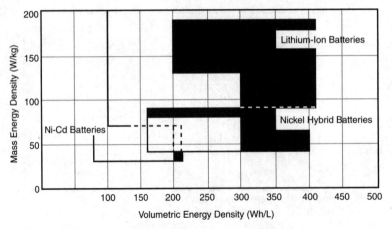

■ **Figure 3.20**   *Energy density chart.*

## Lithium-Ion (Li-Ion)

These are the Ferraris of the battery world.

Because this chemistry is being seen more often in portable products today, it is worth discussing this technology since more and more experimenters will be getting their hands on it. These batteries are the ultimate in lightweight, high-power density, high per-cell voltage, and high cost. They also have excellent shelf life.

At the same time, use of Li-Ion batteries comes at a rather high risk. If not handled properly, these cells can ignite or explode. Every cell has several protection features built in:

- Special vents built into the battery case
- A small circuit board with microcontroller to monitor charge and discharge conditions and open the circuit if these go awry
- A built-in thermistor
- A fuse

The power density of this battery is awesome, as Figure 3.20 shows. Part of what makes the power density so great is the fact that each Li-Ion cell produces 3.7V under load, equivalent to three NiCds. Moreover, lithium batteries have none of the "memory" effects seen in NiCd and NiMH batteries. Lithium batteries exhibit an extremely flat discharge voltage profile throughout the discharge period.

Aside from the risky character of this chemistry, its other limiting factor is that the cells max out at about 1,400-mAH capacity. No doubt manufacturers will continue to increase this number, but for now, that's it. I haven't had the opportunity to read much about the newest lithium technology—lithium polymer. By using the polymer as an electrolyte, the batteries are expected to be much less volatile and can be shaped to most any form. This technology is just hitting the OEM market as of this writing.

Although there are a half-dozen charging schemes for the other chemistries, there is only one for Li-Ion…the carefully *controlled constant voltage/constant current* method. Here are the recommended parameters to keep in mind as the battery is charged. First, the maximum voltage of the cell is 4.2 V. If charging is cut off at 4.1 V, the battery will be about 80 percent charged. Do not exceed 4.2 V. Second, charge at a rate not to exceed 1 C. Panasonic recommends a 0.7 C rate. When cell voltage is 2.9 V or less, begin the charge at a 0.1 C rate. Third, charge at temperatures between 0°C and 45°C. The on-board controller will not allow charging to more than 4.3 V.

Current should be maintained at 1.0 C or less. If discharging a 1,000 mAH battery, limit the current to a maximum of 1,000 mA. Most likely the on-board controller will limit the amount of current that can be drawn from the cell as a form of short-circuit protection. Sometimes, at loads greater than 1 C, even for very short periods of time, the battery will suddenly "go dead" when the controller kicks in. Terminate discharge of the cell at 3.0 V. Below 2.7 V the cell will likely suffer damage. The on-board controller will probably not allow discharge to this level.

There are some remarkable stories about lithium-ion batteries and their volatile behavior. Plants producing these batteries in Japan have burned twice over the past few years. Cellular phones are rumored to have ignited while in use. There is a story of a lithium-ion battery in an Apple laptop igniting at a press conference announcing the new computer. In other words, these batteries have attained legendary status in their short history. Please be very careful with these devices. Here's the list:

- Do not reverse polarity while charging or discharging the cells.
- Do not allow the cell to receive a sharp or sudden impact.
- Do not place the battery in fire or heat the battery.
- Do not carry or store the batteries together with necklaces, hairpins, or other metal objects.
- Do not pierce the battery with nails, strike the battery with a hammer, step on the battery or otherwise subject it to strong impacts or shocks.

- Do not solder directly to the battery!
- Do not expose the battery to water or salt water, or allow the battery to get wet.
- Do not modify or remove the safety and protection devices.
- Do not place the batteries into direct sunlight.
- Do not leave the battery in a car on hot days.

This list was quoted from the Panasonic "Safety Precautions for Lithium Ion Battery Packs." Taking good care of a robot's batteries will assure maximum return on the investment.

## Summary

Technology must enable animatronics figures like MentorBots to become conversationally interactive. Further goals include allowing these characters to become aware of their environment and react accordingly.

These goals are met by combining a multitude of technologies (speech recognition, synthetic interview, discussion engine, audio, vision, and animatronic technologies) into a new medium for interactive entertainment.

The example MentorBot was developed, and "brought to life," through a custom-built animatronic head, off-the-shelf hardware and software, and custom software.

Although a trained actor can be more immersed, entertaining, and interesting than a synthetic being, an animatronic character can do many things that a live actor cannot, such as:

- Work 24 hours a day, 7 days a week
- Take completely nonhuman forms (e.g., talking soup cans)
- Ensure that the show content is always delivered accurately
- Be able to be duplicated and distributed exactly
- Offer an audience the novel experience of interacting with a robot
- Operate completely autonomously

What we are really doing is akin to designing a clever magic trick—if the audience gives in, they think they are talking to a living, breathing character. In other words, the suspension of disbelief. And, if we are successful in suspending the audience's disbelief with technology, that's a cool thing!

Interactive animatronic characters have numerous applications and venues. Here are some prospective utilizations of this technology:

- **Location-based entertainment.** The possibilities are limitless for providing theme parks, amusement parks, themed restaurants, and other location-based entertainment venues with engaging animatronic characters that can be interacted with on a conversational level instead of just being seen and heard.

- **Interactive electronic pet.** Have you ever seen an electronic pet? With the ability to be truly interactive, these pets could be personified and given a genuine character. What if your pet actually told you it was hungry, or thanked you for taking care of it?

- **Embodied personal digital assistant.** Imagine walking into your home or office to be greeted by a full-bodied assistant who can keep you up to date on your schedule, tell you whose birthday you're about to forget, and answer your questions about when and where to be, and even why.

- **Information kiosk.** Airports, theme parks, shopping malls, hotels, and the like all need a concierge or information desk to direct guests. What if an exciting animatronic character could tell you what gate your flight departs from or where a brand new roller coaster is?

- **Restaurant automated waiter.** One day, you may sit down at a restaurant and ask a robot what today's soup is. It could explain the menu, take your order, and even understand service requests. The only question is, how much do you tip?

- **Retail salesperson.** If you've ever worked in retail, you know just how annoying it can be to answer the same questions and direct people to the bathroom over and over again. Interactive animatronic characters can serve as retail clerks for answering questions about merchandise. But if a robot ever says, "You look great in that dress," you may not want to take his word for it.

# 4

# Today's MentorBots

## Stationary Robots

### Kasey the Kinderbot

Kasey the KinderBot (Figure 4.1), a preschool MentorBot, has a warm smile and gentle wave that encourages children to explore all the fun places on Kasey's interactive screen. Children have hours of fun singing, dancing, sharing, and learning with this playful, interactive friend from Fisher-Price. Kasey has a special learning circuit that teaches more than forty developmental skills:

- Academic skills
  - □ Alphabet and phonics
  - □ Numbers and shapes
  - □ Visual discrimination
  - □ Matching and memory
- Social skills
  - □ Manners and politeness
  - □ Friendship and fun
  - □ Emotions and caring
  - □ Taking turns
- Physical skills
  - □ Moving and stretching
  - □ Dancing and singing

■ **Figure 4.1** *User friendly KinderBot.*

☐ Coordination and balance

☐ Fine and gross motor skills

Activities included with the KinderBot contain lessons in reading readiness, early math skills, and physical development:

■ ABC Farm for reading readiness
   ☐ Alphabet song
   ☐ Uppercase and lowercase letters
   ☐ Phonics
   ☐ Word association
   ☐ Food and animal names
■ Counting Kitchen for math readiness
   ☐ Numbers
   ☐ Counting
   ☐ Identifying quantities
   ☐ More or less
   ☐ Shapes and colors
■ Game Room for physical development
   ☐ Dance to the music
   ☐ Sing-along songs
   ☐ Stretch break
   ☐ Move like Kasey
   ☐ Tag Chip game

- Playground for fun learning activities
  - ☐ Trivia game
  - ☐ Dress up with Chip
  - ☐ Match clothes
  - ☐ Same/different animals
  - ☐ Magic hat memory game
- Additional features
  - ☐ Kasey grows right along with a child, so learning as your child develops can be expanded using a variety of software cartridges (sold separately).
  - ☐ There's going to be a lot of giggling, singing, and dancing going on, along with high fives (and other amazing moves!) to reinforce your child's success.
  - ☐ Kasey's eyes and hands move.

## Program Cartridges

The program cartridges are color-coded according to category: red for languages, green for science, blue for math, and yellow for reading. The introductory cartridge that comes with Kasey is purple, but is actually a blank—the introductory program itself is stored directly in Kasey, so that Kasey will always go to that default program if the cartridge slot is empty.

The following initial program cartridges are available:

- Languages: French and Spanish
- Math: Addition & More and Numbers & Counting
- Reading: Focus on Phonics and Words & Sentences
- Science: Living Things and Wonderful World

## Specifications

- Shipping weight in pounds: 5.65
- Product measurement in inches: 15.25 × 6.0 × 10.5
- Batteries not included: You'll need three D alkaline

## Contact:

Fisher-Price, Division of Mattel, Inc.
636 Girard Avenue
East Aurora, New York 14052
Email: fpconaff@fisher-price.com
Phone: (800) 432-5437

## NeCoRo

Created with core sensing, control technology, and artificial intelligence technology, NeCoRo the pet cat realizes natural human-to-robot communication. Because of its ability to react to human movement and express its own emotions, people pour their affection into this robot and feel attached to it as they would a pet. By living with it day after day, the person becomes at ease with the robot as it enriches the person's life.

NeCoRo's synthetic fur gives it a feline appearance and introduces communication in the form of playful, natural exchanges as between a person and a cat. Via internal sensors of touch, sound, sight, and orientation, human action and thoughts can be perceived, and feelings and wants are generated based on internal feelings. Using fifteen actuators inside the body, it behaves in response to its feelings. It will get angry if someone is violent to it, and express satisfaction when stroked, cradled, and treated with lots of love. Based on its own physiological rhythms, it will express its desire to sleep or cuddle. Moreover, through a learning and growth function, while living together day after day, the cat will become attached to its owner and its personality will adjust to the owner. And as it begins to remember the sound of the owner's voice and its own name, it will recognize its name when called out by the owner (Figure 4.2).

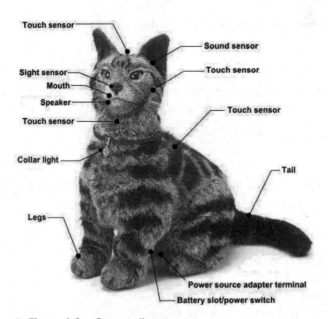

■ **Figure 4.2** *Pet cat diagram.*

### Characteristics

- Responds to human movement/emotions
- Has feelings and desires, and its personality will adjust to its owner
- Remembers its name and acknowledges its name when called
- Synthetic fur makes it feel natural, so it gets a lot of stroking and hugging

### Specifications

- Dimensions
  - Outer dimensions 260 mm × 160 mm × 320 mm (excluding tail)
  - Weight 1.6 kg (when battery is inserted)
  - Fur material Acrylic (gray or brown)
- Sensing and recognition
  - Tactile sensors are embedded in the head, chin, and back to sense strokes and pats
  - A microphone in the head can detect sound and recognize the source of the sound
  - As its name is called over and over, it remembers its name and reacts when called
  - Within its scope of vision, it can perceive the direction of moving objects
  - An internal acceleration sensor allows it to know its position when cradled or spun
- Feeling generation mechanism
  - With Omron's proprietary MaC (mind and consciousness) technology, feelings are generated according to recognition feedback, which is dependent on configurations based on psychological concepts, leading to cognitive decisions and actions determined by these feelings (applicable patent acquired)
  - Feelings of satisfaction, anger, and uneasiness generated based on recognition feedback
  - Desires to sleep or be cuddled generated according to physiological rhythms
  - Via a learning function, personality traits such as selfishness and the need for attention will change in response to the owner
  - Via a growth function, expressive patterns in reaction to the owner will increase
- Actuators
  - Four legs (each with 2 degrees of freedom), tail with two degrees of freedom, neck with 2 degrees of freedom, and eye-

lids, ears, and a mouth each with 1 degree of freedom, for a total of 15 degrees of freedom.
- □ Equipped with many movement patterns, a voice speaker and 48 different vocalized cat sounds
- ■ Power supply
  - □ Replaceable nickel hydrogen battery (Ni-MH)
  - □ Battery is good for 1.5 hours of operation (battery takes 2 hours to fully charged)
- ■ Accessories
  - □ Battery pack/special charger/instruction manual/ brush
  - □ Internal sensors and primary operational parts

### Contact

Christopher Udell, Corp. Communications Headquarters, Tokyo
TEL: 81-3-3436-7139, christopher_udell@omron.co.jp
NeCoRo Customer Center:
Tel.0120-066-792 or Omron's Web site at http://www.omron.com/

## CosmoBot/JesterBot

Most children with disabilities have one or more therapeutic needs that are not being met. Very few products make therapy engaging or exciting to motivate children to meet their functional goals. Nor are there products that automatically track the child's progress to justify or decide on appropriate interventions. Through innovative applications of technology, we are exploring the development of products, methods, and systems that will enable children with disabilities through interactive play.

AnthroTronix is developing telerehabilitation tools (Figure 4.3) to motivate and integrate therapy, learning, and play. These technologies are developed in conjunction with therapists, educators, parents, and children. This research has been funded by a Small Business Innovation Research grant from the National Science Foundation (NSF SBIR), the Maryland Industrial Partnerships, and the Rehabilitation Engineering Research Center (RERC) on Telerehabilitation (for more info on the RERC on Telerehabilitation, see: www.telerehab.cua.edu).

The nearly 3 million children with disabilities (ages 3–14) are the target population of this development project. The child experiences enjoyment and self-determination and increased motivation for engaging in the learning or therapeutic activities:

■ **Figure 4.3** *CosmoBot and JesterBot.*

- **Promotes communication skills.** Programming the robot to respond to the child's voice revealed a key motivational aspect of the GIRT system. In addition, these functions facilitated interactive storytelling for speech-language development, a key goal for all children.
- **Creates a flexible, usable system.** The robot serves a variety of functions—mimicking the child's movement, leading the child through exercises, and playing back previously recorded scenarios. The software interface must be easy to use for all parents, therapists, and children.

AnthroTronix has developed CosmoBot™ and its predecessor, JesterBot™, robotic toolkits designed for clinical rehabilitation and special education. CosmoBot™ is controlled by body movements, voice activation, or Mission Control, an interactive control station. CosmoBot™ and JesterBot™ are being tested in clinical and educational settings throughout the development process. By incorporating children into the preliminary stages of testing, we increase the efficacy of the prototype and explore children's responses to stimuli. The data also led to developing methodologies for therapists and educators to follow in providing assistance to children with special needs.

# MDM Robot

## The MDM (Million Dollar Machine®) Program

MDM (Figure 4.4) gives children the knowledge and motivation they need to achieve their personal best in life. MDM gives children a new perspective of themselves and then teaches a variety of useful skills to help them interact with other people and the world around them:

■ **Figure 4.4**  *MDM robot.*

- High-tech, high-touch personality captivates children of all ages
- Every custom scripted educational assembly includes school, teachers, and students
- LED sign displays educational messages
- The robot teacher presents each special assembly live in the school
- Dispenses printed materials
- Self-contained 200 W digital music and sound effects require no special staging or extra set-up

What "life skills" does MDM teach?

The age-specific study materials and optional school assembly presentations cover five key topics:

- Self awareness
- Interpersonal skills

- Decision-making
- Drug awareness
- Refusal skills and earth skills

### Character Education

MDM is effective for all children in Grades K–6 (ages 5–12). The lesson plans are designed to be easy for teachers and parents to use with all children in this age group. Who is the "Million Dollar Machine?" MDM teaches children that their bodies are the real, irreplaceable "million dollar machines" for which the program is named.

### Quick Facts on the MDM Program

- MDM motivates children to achieve their personal best by teaching responsibility, decision-making, and positive social skills.
- MDM includes specific health and assertiveness skills that inspire children to protect themselves from cigarettes, alcohol, drugs, violence, and other threats.
- MDM's effectiveness is validated by five independent scientific studies.
- More than 2,500,000 U.S. children have benefited from MDM since its introduction in 1986.

### Contact

LifeSkills4Kids.com
12 Phillips Road, Hainesport, NJ 08036
TEL: 609 261-2162 FAX 609 261-1512

# Mobile Robots (Flat Surfaces)

## Sony's AIBO-31L

The newest member of the AIBO family, the ERS-31L, has a unique design resembling a bulldog (Figure 4.5). The ERS-31L offers all of the features of the AIBO LM series at an affordable price. The ERS-31L is 11 inches tall and 7 inches wide and comes in caramel brown. It recognizes and responds to seventy-five voice commands. Lights, touch sensors, and a new tail switch make for clear communication with AIBO. With the new software AIBO Pal Special Edition, the ERS-31L will further entertain with 200 dance moves. Other characteristics include a special wake-up dance when AIBO is powered on. Also, this

129

curious AIBO walks around looking for things of interest and snapping photographs—you will be surprised to see how AIBO sees its surroundings. Features include:

- Mode indicator
- Horn light
- Head
- Stereo microphone
- Distance sensor
- Color camera
- Speaker
- Paw switches
- Back light
- Pause button
- Tail switch

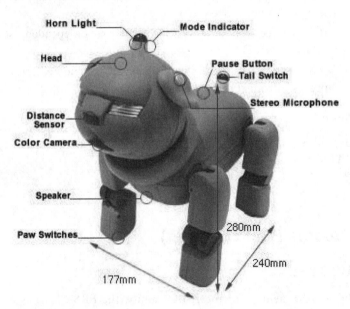

- **Figure 4.5**  *AIBO ERS-31L.*

## Specifications

AIBO ERS-31L, ERS-311C, and ERS-312H

- CPU—64bit RISC processor
- Main memory—32MB SDRAM

- Supplied application media—Memory stick for AIBO
- Movable parts—Head: 3 degrees of freedome (DoF) (PAN × 1 TILT × 2) Leg: DoF × 4, 15 degrees of freedom in total
- Inputs—Charging terminal
- Input switches—Volume switch
- Visual input—100,000 pixel CMOS image sensor
- Audio input—Stereo microphone
- Audio output—Speaker
- Internal sensors—Infrared distance sensor and acceleration sensor, switches on head, tail (ibi-axis analog), and paws (four), inclination sensor and vibration sensor
- Power consumption—Approximately 5 W
- Operable duration—Approximately 2.5 hours
- Dimensions—Approximately 177 × 280 × 240 mm
- Mass—Approximately 1.5 kg (includes battery pack and memory stick)
- Operating temperature—5°C to 35°C
- Operating humidity—10 to 80 percent
- Storage humidity—10 to 90 percent

**Contact**

Sony Corporation
http://www.us.aibo.com/

## Sony's AIBO ERS-210A

The ERS-210A (Figure 4.6) is an upgraded high-speed version of the ERS-210, with a new advanced central processing unit. The ERS-210A is quick, capable of running enhanced software applications, and exhibits greater functionality. For example, AIBO will respond faster to commands when using software such as AIBO Navigator 2 via Wireless LAN. (Some AIBO-ware software are not optimized for the ERS-210A.)

AIBO, with its own emotions and instincts, learns and grows through human interaction and from its environment. Owners have a variety of software applications to choose from to enhance their AIBO experience.

■ **Figure 4.6** *AIBO ERS-210A.*

## Features

- IBO memory stick
- Advanced central processing unit
  - ☐ CPU clock speed of 384 MHz, doubling the previous speed of 192 MHz
  - ☐ Mouth, 1 degree of freedom
  - ☐ Head, 3 degrees of freedom
  - ☐ Leg, 3 degrees of freedom × 4 legs
  - ☐ Ear, 1 degree of freedom × 2
  - ☐ Tail, 2 degrees of freedom
  - ☐ Total 20 degrees of freedom
- Built-in sensors
  - ☐ Temperature sensor
  - ☐ Infrared distance sensor
  - ☐ Acceleration sensor pressure sensors (head, back, chin, paws)
  - ☐ Vibration sensor
  - ☐ PC card slot Type2 in/out
  - ☐ Memory stick slot in/out
  - ☐ AC IN power supply connector input
- Image input CMOS image sensor
- Audio output miniature speaker
- LCD display

- ☐ Time
- ☐ Volume
- ☐ Battery condition
- Built-in clock with date and time display
- Power consumption approximately 9 W (standard operation in autonomous mode)
- Operating time approximately 1.5 hours (standard operation in autonomous mode)
- Operating temperature 41°F to 95°F (5°C to 35°C)
- Operating humidity 10 to 80 percent
- Dimensions and weight
  - ☐ Size (W × H × L) 6.06" (W) × 10.47" (H) × 10.79" (L)
  - ☐ Weight 3.3 lbs. (1.5 kg)
- Supplied accessories
  - ☐ AC adapter
  - ☐ Lithium ion battery pack
  - ☐ Ball
  - ☐ Documentation

**Contact**

Sony Corporation
http://www.us.aibo.com/

## RB5X Classroom Robot

The RB5X Educational Assistant (Figure 4.7) is an educational tool capable of increasing a student's attention, comprehension, and motivation for learning in mathematics, science, language arts, social studies, and computer technology at all grade levels. RB5X has been shown to help children increase their math and language skills, personal motivation, and intrigue for learning.

The RB5X has proven to be a powerful teaching tool. When teaching the RB5X math, language arts, or social studies, students not only demonstrate their knowledge and identify their learning gaps, they develop self-esteem and problem solving skills, critical for success in today's world. Through this transformation, students take responsibility for their educational experiences and gain a lifelong motivation for learning. The RB5X interacts with a class in ways that a computer or robotic arm alone simply cannot do.

■ **Figure 4.7**    *RB5X robot.*

The RB5X incorporates infrared sensing, ultrasound sonar, remote audio/video transmission, eight sensors/bumpers, a voice synthesizer, and a five-axis armature that can lift a full pound. The RB5X can play interactive games with up to eight people. Programs can be written and downloaded from any computer, Apple Macintosh, DOS, Windows, or Linux/UNIX.

### Contact

General Robotics
760 South Youngfield Court
Lakewood, Colorado 80228-2813
http://www.generalrobotics.com
(800) 422-4265

## PeopleBot

PeopleBot™ (Figure 4.8) provides a base for service or performance robots. PeopleBot offers a gripper, table-sensing IRs and a precise pan-tilt camera for sensing and grasping objects on tables. PeopleBot is available as a complete system, with components and accessories that vary by model, or as a developer base. Every PeopleBot includes:

- Body with hinged battery access door
- Vertical member(s) and upper platform
- Three batteries with charger

■ **Figure 4.8** *PeopleBot Guide.*

- Two wheels and one caster
- Two motors with encoders
- Front sonar rings, upper and lower
- Lower rear sonar
- Fixed IR for table
- Other mid-level avoidance
- Microcontroller
- Sonar board
- Motor power board
- Complete software package
- Operations manual

PeopleBot has the ability to:

- Play sound files or synthesized speech
- Listen for phrases or sounds it recognizes
- Respond to requests or conditions it senses
- Navigate without running over toes or into furniture
- Find and fetch objects it recognizes
- Follow colors
- Transmit video images to surveillance monitors

- Connect to PCs via the Internet or LAN
- Run autonomously

## Technical Specifications

The Performance PeopleBot model's body is made of aluminum. PeopleBot has two 19 cm diameter drive wheels with 36:1 gear ratios that have power to carry a 13 kg payload. Their 500-tick encoders provide <1 percent dead reckoning error. These differential drive platforms are highly holonomic and can rotate in place moving both wheels, or can swing around a stationary wheel in a circle of 32 cm radius. A rear caster balances the robots.

PeopleBot can climb a 5 percent grade and sills of 1.5 cm. On flat floor, it can move safely at speeds of .8 mps; faster speeds are possible, but not recommended. At slower speeds it can carry payloads up to 13 kg. Payloads include additional batteries and all accessories and must be balanced appropriately for effective operation of the robot.

In addition to motor encoders, all PeopleBot robots include twenty-four ultrasonic transducer (range-finding sonar) sensors arranged to provide 360-degree forward coverage. The sonar read ranges from 15 cm to approximately 7 m. Some robots also include rear sonar.

PeopleBot robot's hinged battery door makes hot-swapping batteries simple, though a bare PeopleBot™ base can run for 18 to 24 hours on three fully charged batteries. With a high-capacity charger, recharging time is only 2.4 hours.

The PeopleBot robot's easily removable nose allows quick access to any optional embedded computer for addition of up to three PC104+ cards. PeopleBot includes a Siemens C166-based microcontroller. The microcontroller, has eight digin and eight digout plus one dedicated A/D port; four digin can be reconfigured as A/D in; four digout can be reconfigured to PWM outputs. This user I/O is integrated into the packet structure, accessible through ARIA and Saphira.

PeopleBot also includes a fixed IR with a range of 50 mm to 1000 mm to sense the underside of tables. It points up and slightly forward from the nose of the robot base. The IR beam is collimated using special lensing to allow reliable long-range reflective sensing.

Performance PeopleBot includes a vertical 1 DoF beam device with horizontal 1 DoF gripper. The gripper has a maximum 12 cm (4.75 in.) spread between fingers. The combined beam/gripper mechanism has a payload capacity of 1 kg (2.2 lbs.)

A small proprietary P2OS transfers sonar readings, motor encoder information, and other I/O via packets to the PC client and returns control commands. ARIA supplies the developer interface, for use under RedHat Linux with Motif or under Windows using a C/C++ compiler. Saphira 8 offers localization and gradient navigation.

PeopleBot is designed for use by seasoned professionals and can be programmed in C or C++. PeopleBot is ideal for prototyping, research, or applications such as:

- Tour guides
- Waiters
- Messengers
- Monitors and guards
- Trade shows and exhibitions
- Performances
- Education
- Research
- Cooperative tasks

PeopleBot runs indoors on flat floors. It can traverse low sills and household power cords. With upper and lower sensing, it will turn away from nearly all obstacles. Performance PeopleBot also has the ability to sense tabletops and move its gripper into place for picking up objects. PeopleBot can run five days a week for six hours a day for years without maintenance. More intensive use may require regular factory maintenance, which is available by contract. PeopleBot bases may communicate with each other via Ethernet and cooperate in teams.

### Contact

ActivMedia Robotics, LLC
19 Columbia Drive
Amherst, NH 03031 US
Ph: +1 603-881-7960 Fax: -3818
Email: robots@activmedia.com
URL: http://www.MobileRobots.com

## Wakamaru

A childlike robot that combines the roles of nurse, companion, and security guard is to go on the market to help the growing ranks of elderly Japanese with no one to look after them (Figure 4.9).

■ **Figure 4.9** *Wakamaru robot nurse.*

The Wakamaru robot can patrol a house 24 hours a day, alerting family, hospitals, and security firms if it perceives a problem. It will call relatives if its owner collapses or fails to get out of the bath.

The technology has gained nationwide publicity in Japan amid increasing concern over how to look after the ever-growing number of old people. Wakamaru has its daily rhythm of life, speaks to people, and lives with family members.

### Features

- **Wakamuru lives with family members.** The robot lives in accordance with the stored day's schedule and lives together with the owner by updating the schedule based on contact with the owner.
- **It speaks spontaneously in response to its family members.** Not only does the robot respond to actions from people, but it also speaks to family members based on the information obtained from the contact with the family.
- **It has its own role in a family.** The robot connects itself spontaneously to the network to provide necessary information for daily life, looks after the house while the family is absent, watches for unusual cases, and fits conveniently into the life of family members.

### Specifications

- Dimensions
  - □ Height—100 cm
  - □ Width—45 cm

- Weight—Approximately 30 kg
- Maximum moving speed—1 km /hour
- Joints
  - Neck—3 DoF
  - Arm—4 DoF × 2 arms
  - Moving parts—2 DoF wheels
- Drive method—DC servo motors
- Power source
  - Battery—Operates continuously for maximum 2 hours
  - Charging—Automatically charged at the charging station
  - CPU—Multiprocessor configuration
- Control section
  - Operating system—Linux
  - Robot control architecture—Serial bus connection distributed processing, omni-direction camera
  - Sight—Front camera
  - Sense of hearing—Directional microphone and three nondirectional microphones
- I/O section
  - Touch/Force sensing—Shoulder touch sensor, neck touch sensor, wrist load detection sensor
  - Voice—Speaker
- Irregular floor mobility—10 mm protrusion maximum
- Sensor
  - Own position measurement
  - Infrared ray obstacle detection
  - Ultrasonic obstacle detection
  - Inclination detection
  - Collision detection
- Autonomous mobility
  - Autonomous movement to map registered location in a house
  - Autonomous "following" and "finding" a person
- Obstacle avoidance  Autonomous avoidance and rerouting
- Charging
  - Autonomous movement to the charging station and charges itself, when the charged power is lowered
  - While it is charging, it uses controllers and sensors to maintain connection to the Internet

- Human detection (Sensor combined processing)
  - Detection of moving persons
  - Face extraction
  - Thermal source detection
  - Sound source direction detection
- Individual recognition—Detects face characteristics and recognizes two owners and eight other persons
- Voice recognition—Recognizes approximately 10,000 words
- Speech synthesis—Text-to-speech (TTS); you can change the speed of talking and the tone of voice
- Equipment configuration
  - Robot main unit
  - Charging station
  - Wireless broadband router
- Environmental conditions of use
  - Indoor barrier-free floor (space of a general house)
  - Normal temperature
  - Daylight or illumination
  - Continuous Internet connection
- Safety for people
  - Consists of motors of 80W or less in compliance with the Japanese Domestic Material Safety Standard
  - Complies with the Japanese domestic standards for home appliances
  - Joint structure to prevent winding or pinching
  - Arm: Full axis servo level collision detection (torque monitoring)
  - Hand: A collision sensor is installed in the wrist for collision detection

Mitsubishi Heavy Industries, which developed Wakamaru, plans to sell the robot beginning in 2004 for £5,000 to £6,000. Wakamaru speaks with the voice of either a boy or a girl and is also designed to provide companionship. It can be set to remind forgetful people when it is time to take medicine, eat, and sleep.

This is the first household robot able to hold simple conversations based on a vocabulary of around 10,000 words. It can not only speak but also can understand answers and react accordingly. It will ask "Are you all right?" if its owner does not move for some time. If the answer is no, or there is no answer, it will telephone preset numbers, transmitting images and functioning as a speakerphone. Wakamaru will notify a security firm if there is a loud bang or if an unknown person enters the house while the owner is out or asleep. It can recognize up to ten faces. But, like most robots, it cannot climb stairs.

### Contact

Mitsubishi Heavy Industries, Ltd. http://www.mhi.co.jp

## Mobile Robots (Bipedal)

### PINO Humanoid Robot

What reasons are there for designing the humanoid robot? Existing humanoid robot research centers on the development of either a humanoid machine from a mechanical engineering approach or, conversely, an analytic machine by which the mechanisms of thought or intelligence can be simulated and put into effect by a freely moving body reacting to diverse sensory information.

However disparate the means by which humanoid robot research has evolved, both approaches are concerned with the human form as representative of its mechanical features. Aesthetics, we believe, will play an even larger role in the design requirements of the robot, playing a pivotal role in establishing harmonious coexistence between the consumer and the product. Accordingly, research that employs an element of aesthetics is inseparable from the robot's primary mechanical functions

### Exterior Design

PINO (Figure 4.10), the humanoid robot was developed in anticipation of the RoboCup Humanoid League. The importance of exterior design is connected first and foremost to the protection of its inside system in much the same way a car or computer is shielded from contact. Providing a mere protective shield to reduce the risk of damage to its inner systems during performance does not sufficiently express research direction, which aims to express the role of the humanoid robot in future society.

Before PINO went into development, size turned out to be a specific element of design research that was carried out simultaneously during primary research into walking functions.

The scale of a fully grown adult posed a threatening presence and being less a companion than a cumbersome and overpowering mechanical object. As for form, it was deemed necessary to design PINO's proportions as recognizably human as possible; deviating too far from the instantly recognizable form of a human child could cause it to be seen as an altogether different object.

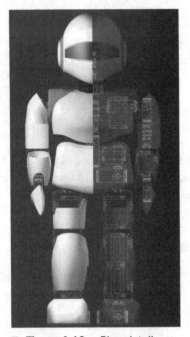

■ **Figure 4.10** *Pino detail.*

### The Origin of the PINO Name

Before starting the design sketches, rearchers looked for a universal element in the representation of the human form. Folkloric marionettes exhibited both obvious beauty, but a mournful aspiration towards perfection. The marionette, with its mechanisms to facilitate movement and expression, provided the ideal framework to which PINO could be adapted. Pinocchio seemed an apt metaphor for our search for human qualities within the mechanical structures of our creation. In his gestation, PINO symbolically expresses not only our desires but humankind's frail, uncertain steps towards growth and the true meaning of the word *human*.

### Specifications

- Dimension
    - ☐ Height 75 cm

- □ Width between both shoulders 32 cm
- □ Width of hip 20 cm
- □ Length of leg 30 cm
- ■ Weight 4.5 kg
- ■ Degree of Freedom
  - □ 26 DoFs
  - □ Neck 2 DoFs
  - □ Backbone 2 DoFs
  - □ Arms 5 DoFs in each
  - □ Legs 6 DoFs in each
- ■ Actuator
  - □ Servo module (torque 20 kg-cm), 14
  - □ Servo module (torque 9 kg-cm), 12
- ■ Sensor
  - □ Vision sensor (CMOS sensor), 1
  - □ Force sensor (FSR), 8
  - □ Joint angle sensor (Potentiometer), 26
- ■ Host computer
  - □ PC (Pentium III 733 MHz)
  - □ Memory 512 MB
- ■ Operating system
  - □ Real-time Linux 6.2 Kernel 2.2.14-5.0
  - □ RT-Linux 2.2
- ■ Development tool
  - □ gcc (C language)
  - □ ALTERA Max+PLUS II BASELINE
  - □ YCSH (C language)
- ■ Robot controller
  - □ SH2 (HD64F7050 20 MHz)
  - □ CPLD (EPF10K30ARC240-3)
- ■ AD board
  - □ Contec AD12-64 (PCI)
- ■ Frame grabber
  - □ I/O DATA GV-VCP2/PCI
- ■ Cable
  - □ Small gauge coaxial cable coaxial cable × 50, power (7A)
- ■ Power resource

- ☐ EWS300-6
- ☐ Input: AC 100-120 V 8A, AC 200-240 V 4 A, 50/60 Hz, 410 W
- ☐ Output: DC 6 V 50 A
- ■ Exterior
  - ☐ Laser beam lithography (30 parts)
- ■ Structural material
  - ☐ Duralumin
  - ☐ ABS regin

### Contact

ZMP, Ltd.
PINO Open Platform
Mita one Chome 3rd
39 Katsuta Building 7th Floor
Minato-ku, Tokyo 108-0073, Japan
Tel: +81-03-5765-6567 Fax: +81-03-5765-6568
E-mail: info@pinodx.com

**144**

## Summary

A window of opportunity is opening. For 30 years, the effort to create robots that can reliably and competently negotiate novel locales on their own and interact with humans has been unsuccessful. In consequence, a mass market for mobile robots (other than toys) has failed to materialize. Computer power and the necessary techniques to make true autonomous mobility possible are finally becoming available. Within a decade they will be widespread. The first major products of this new industry are designed for marketing in the next several years. Today's MentorBots fall into this category.

Since 1980, the development of interactive machines has progressed to the point of conversational robots. Since 1992, programs have been initiated to process data from robot sensors, especially stereoscopic cameras. These two technologies form the core of MentorBot functionality: improved preliminary processing of images, automatic optimization, route planning, robot localization within maps, object classification, and many application-specific topics. These features might be addressed early in commercial work, but are also suitable for university research now. From the research point of view, a modest investment in this direction could have a disproportionately high payback by sparking vigorous, self-sustaining growth. A lively industry would be fertile soil for promising robotics research that today is often lost for lack of a sustaining context. Many precedents exist in the computer industry: Computers, integrated circuits, and the Internet, once research initiatives, now grow vigorously in the open market. We judge robotics to be on the verge of an analogous flowering. Keeping the work in an open research setting as long as possible will maximize its availability to others.

The concept of using self-guiding mobile robots for child rearing, companionship, domestic help, and as senior caregivers has existed for most of the twentieth century, but has been realized in only modest ways. A few tens of thousands of automatically guided vehicles (AGVs) are at work in factories, warehouses, and other institutional locations. Many are guided by buried signal-emitting wires that define their routes and use switches to signal endpoints and collisions, a technique developed in the 1960s. More advanced models, made possible by the advent of microprocessors in the 1980s, use more flexible cues, such as optical markings on a tiled floor. The most advanced machines, manufactured in the late 1980s and 1990s, are guided by navigational markers, such as retro reflective targets at strategic locations, and by specialized site-specific programming that exploits existing features, like walls.

The newer systems can be installed with less physical effort, but all require the services of a specialist to program the initial setup, and for layout and route changes. The expense, inconvenience, inflexibility, and lack of independence stemming from the elaborate setup greatly reduces the potential market for these systems. Only very stable and high-value routes are candidates, and the high cost reduces any economic advantage over human-guided vehicles. Fully autonomous robots, which could be simply shown or led through their paces by nonspecialists, then trusted to execute their tasks reliably, would have a far greater market.

A generation of fully autonomous mobile robots, which navigate without route preparation, has emerged in research laboratories worldwide in the 1990s, made possible by more powerful microprocessors. The majority of these use sonar or laser rangefinders to build coarse 2-D maps of their surroundings, from which they locate themselves relative to their surroundings and plan paths between destinations.

Mobile robot research, ongoing since the 1970s, has investigated sparse 3-D models and dense 2-D maps, using camera, sonar, and laser sensors. In the last five years, taking advantage of the most powerful microprocessors, we have developed efficient programs that maintain

3-D volumetric maps of a robot's surroundings, containing about a thousand times the information of 2-D maps. As a bonus, the 3-D maps produced can be used in the robot to recognize doors and furniture-sized objects in the surroundings.

A projected first commercial product from mobile robot research is a basketball-sized "navigation head" to be retrofitted on existing robots, providing them with full autonomy. It would contain 360-degree stereoscopic cameras and other sensors, an inexpensive inertial navigation system to inform it about small motions without depending on accurate odometer measurement from robot wheels, 1,000 MIPS of computational power, software for basic navigation, software hooks for applications-specific programming, and hardware interfaces for vehicle controls. Offered as an OEM product to existing AGV manufacturers and others, it could quite possibly expand the market for AGVs.

A possible first product with mass-market potential is a small robot vacuum cleaner, which can reliably and systematically keep clean designated rooms in a home following a simple introduction to the location.

The simple vacuum cleaner may be followed by larger and smarter utility robots with dusting arms. In subsequent products, arms may become stronger and more sensitive, to clean other surfaces. Mobile robots with dexterous arms, vision, and touch sensors will be able to do various tasks. With proper programming and minor accessories, such machines could pick up clutter, retrieve and deliver articles, take inventory, guard homes, open doors, mow lawns, or play games. New applications will expand the market and spur further advancements in acuity, precision, strength, reach, dexterity, or processing power. Capability, numbers sold, engineering and manufacturing quality, and cost effectiveness should increase in a mutually reinforcing spiral. Ultimately, the process could result in "universal robots" that can be do many different tasks, orchestrated by application-specific programs.

# Build Your Own
# MentorBot

# 5

BESIDES A HANDFUL OF CHILDREN'S MENTORBOTS LIKE AIBO ERS-31L, CosmoBot, and Kasey the Kinderbot (see Figure 5.1), Mentorbots generally cost over $1,000.

In fact, all the MentorBots under $3,000 are either small animals or child-oriented, like AIBO ERS 210A, RB5X, and NeCoRo (see Figure 5.2).

If you want an adult MentorBot under $3,000, and have the skills to build one, this may be your chance. This chapter contains plans for two adult-sized MentorBots, one under $1,000 and another under $3,000.

## Overview

This chapter is written for those who cannot wait for their MentorBots to reach the market or who want the adventure of building their own. Construction is compartmentalized into prioritized modules and addressed in the order the modules should be built. The "conversation module" is the highest priority, followed by the "vision module." Since a head and face are needed to make the first two modules effective, the "head module" and "face modules" are built next. A "talking head" may be enough for some builders, but most will also want a "torso module" to provide a place to set the "head module" and to hold the electronics. The majority of builders will not be satisfied with a stationary MentorBot, so the "mobile base module" is needed to provide motion and a housing for the power source. Some builders may want

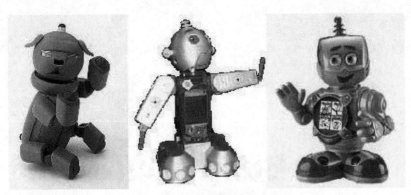

■ **Figure 5.1**  *MentorBots under $1,000.*

■ **Figure 5.2**  *MentorBots $1,000 to $3,000.*

to communicate with their MentorBot via Internet or phone, so a "communications module" will be needed. A few builders will want their MentorBot to pick up and deliver things to them, so the "arm–hand module" should be addressed. Finally, the software to make the MentorBot function needs attention, so the "experimenter interface" comes last. A list of the prioritized modules includes:

- Sound location module
- Speech recognition module
- Speech analysis module
- Speech synthesis module
- Binocular vision module
- Head feature module
- Facial animator
- Torso module
- Joint control module

- Ultrasonic ranging module
- Wheel control module
- Arm–hand module
- Network monitor/control
- Experimenter interface module

First, you need a way to make all these features mobile, so the first order of business is to build a mobile platform for your MentorBot. You could start from scratch and reinvent the wheel, but that would take time away from the major goal of making the companion MentorBot, so a kit is being provided to speed up the construction of the mobile platform.

## MentorBot Platform Kit

All stable structures are built on a solid foundation. The MentorBot Platform Kit is no exception, and its foundation is the motor base (Figure 5.3).

■ **Figure 5.3**   *Robot motor base.*

### Basic Motor Base

The motor base is 40 cm (16 in) in diameter. A pair of 20 in-lb motors, two 15 cm (6 in) wheels, and two 7.5 cm (3 in) casters come with the base.

### Motor Base Specs

- The maximum recommended payload is 35 lbs (13.6 kgs)
- The base has dual 12-V, 20 in-lb torque drive motors; (the max speed is 1 foot per second under full load)—a gear change can double the speed
- Two optic detectors are included and can be used for simple pulse encoders
- Drive wheels are 15 cm (6 in) in diameter
- The casters are 7.5 cm (3 in) in diameter
- The two casters balance the base

## Drive Motor with Encoder Wheel

The drive motor is shown in Figure 5.4.

### Motor Specs

- 12 VDC permanent magnet
- Parallel shaft gearmotor
- 25 rpm 1.30 amps
- Overhung load 18 lbs
- Gear ratio 270:1
- Mounting brackets

The included encoder patterns should be glued to the wheel. The optics can be mounted on top of the drive motor gearboxes using electrical tape.

■ **Figure 5.4**  *Drive motor, gearbox, wheel and swivel caster.*

## Swivel Caster (3 inch)

The heavy duty, swivel casters front and rear are used to balance the platform on the drive wheels (Figure 5.4). The casters are on adjustable mounts, so the amount of teeter-totter rocking can be minimized.

## HC11 Controller

The HC11 processor has a C, Basic, 32K RAM, and 32K EEPROM software example disk (Figure 5.5). It has an easy-to-hook-up terminal strip for I/O and power connections, and includes an expandable back plane with two open slots. The MC motor driver also plugs directly into the back plane. Features of the HC11 controller include:

- Eight 8-bit analog-to-digital converter channels
- EEPROM software data protection switch
- Expandable back plane, up to four cards; 26-pin header
- PC programming software including Basic, C, and Monitor
- Small size, 5 cm × 5 cm cards
- Twenty-two programmable I/O lines
- Three programmable interrupts
- Counter/timer
- Serial cable, HC11 manual
- No EPROM burner or additional hardware is required to program the controller

■ **Figure 5.5** *HC11 controller and accessories.*

## MC TI SN754410 Dual Motor Drive Board

Both motor speed and direction are controllable. Motor speed is variable by adjusting the duty cycle of the PWM input. The MC TI dual motor drive is made to plug into and work with the HC11 MicroCore controller, all of which are included in the basic MentorBot kit.

### Motor Drive Specs

- Requires only two digital I/O lines to control each DC motor
- Simple interface reduces microcontroller I/O count
- Drives two DC motors at up to 1 A per channel
- Plugs into the HC11 backplane
- Completely assembled

■ **Figure 5.6**   *MC motor drive and MC sonar/encoder driver boards.*

## MC Sonar/Encoder Driver Board

The sonar/encoder driver board provides the high-current 5-V DC power supply required for the Polaroid 6500 sonar unit. It also provides sonar multiplexing, which allows one input to read all three sonar units. The multiplexing reduces the required I/O used for three sonar units to one input and three outputs. The board also provides power and acts as a biasing resistor network, required for the simple pulse encoder optics included with all the basic mobile robot packages. This board drives up to three sonar units and two encoder optics, and plugs directly into the HC11 MC back plane.

## Polaroid 6500 Sonar Ranging Unit

Specifications include:

- Accurate sonar ranging from 6 inches to 35 feet
- Drives 50-kHz electrostatic transducer with no additional interface
- Uses TI TL851 and Polaroid 614906 sonar ranging integrated circuits
- Accurate clock output for external use
- Variable gain control potentiometer
- Multiple measurement capability
- Convenient terminal connector
- Selective echo exclusion
- Socketed digital chip
- TTL-compatible

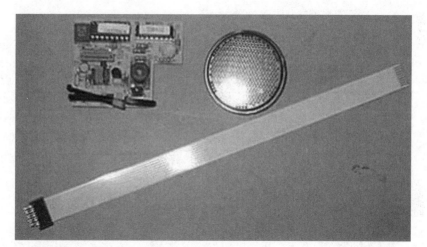

153

■ **Figure 5.7**  *Polaroid 6500 sonar ranging unit.*

The 6500 Series is an economical sonar ranging module that can driven all Polaroid electrostatic transducers with no additional interface. Typical absolute accuracy is ±1 percent of the reading over the entire range.

## Magnetic Digital Compass

It is extremely useful to tell what direction the robot is going as related to the world and not just the direction of the wheels. This compass was designed specifically for use in robots to aid navigation (Figure 5.8). The aim was to produce a unique number to represent the direction the robot is facing. The compass uses the Philips KMZ51 magnetic field sensor, which is sensitive enough to detect the earth's magnetic field. The output from sensors mounted at right angles to each other is used to compute the direction of the horizontal component of the earth's magnetic field.

Specifications for the compass include:

- Voltage 5 V, Current 20 mA typical
- Size 32 mm × 35 mm; weight 0.03 oz
- Resolution 0.1 degree; accuracy 3–4 degrees approximately (after calibration)
- Output 1 timing pulse 1 ms to 37 ms in 0.1 ms increments
- Output 2 I2C interface, 0–255 and 0–3599
- SCL speed up to 1MHz

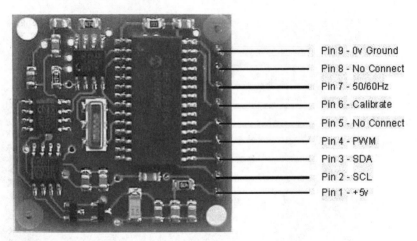

Pin 9 - 0v Ground
Pin 8 - No Connect
Pin 7 - 50/60Hz
Pin 6 - Calibrate
Pin 5 - No Connect
Pin 4 - PWM
Pin 3 - SDA
Pin 2 - SCL
Pin 1 - +5v

■ **Figure 5.8**  *Digital magnetic compass.*

## Assembled MentorBot Platform

The kit includes two 40-cm decks with tie rods and spacers, bringing the assembled platform to a height of 44 cm (17 in). Holders for 8 'D' batteries are provided, which is sufficient to run the equipment for several hours. The footstool-level platform is ideal for serving or transporting items. The MentorBot Platform (Figure 5.9) is capable of navigating independently while avoiding obstacles and going point to point.

■ **Figure 5.9**  *Assembled MentorBot platform.*

## Basic MentorBot Kit

A new mobile robot concept, using a notebook computer (either PC or Mac) to act as the brains, has been employed for the basic MentorBot. This allows the builder to construct a robot for his (or her) own needs while using familiar controls. A kit should appeal to more hobbyists than starting from scratch to build a MentorBot.

### Talk To Your MentorBot

Robots don't have ears; they use microphones instead. But to hear and understand what a human says requires a special array microphone that picks up the human voice clearly, no matter what direction it is in relation to the robot. Robots don't have a larynx either, so they rely on speakers to be heard. Stereo speakers provide the best arrangement for humans to hear (Figure 5.10).

If the robot is more than a few feet away or in another room, even the array microphone has a problem understanding the speaker. The best plan then is a wireless headset which uses a small boom microphone to speak into and an earpiece to listen to. The BlueTooth wireless scheme

■ **Figure 5.10**   *Array microphone and speakers (Option A).*

to communicate with a robot because it is economical and works up to 33 feet from the robot.

A sleek, over-the-ear headset incorporates the latest BlueTooth wireless technology (Figure 5.11). Volume control is at your fingertips, and the device weighs less than an ounce. Simple operation automatically links the headset to your BlueTooth-compatible device for encrypted calls up to 33 feet away.

■ **Figure 5.11**   *BlueTooth headset and adapter (Option B).*

It will give you up to 4 hours of talk time and 100 hours of standby on a charge, and it fully recharges in just an hour. Just slip the headset into its own holder/charger/adapter and plug it into an AC outlet. The holder/charger/adapter has a built-in belt clip, making it easy to take along wherever you go.

### Talk Hands Free, Cord Free

- Uses BlueTooth technology
- Frequency band: 2.45 GHz
- Range up to 30 feet
- Lightweight (1 oz)
- Talk time: 4 hours
- Charge time: 2 hours

- Standby time: 100 hours

**Specifications**

- Interface: USB 1.1
- Plugs into notebook
- Connection: Wireless
- Network standard: V1.1: 1 Mbps
- Computers supported: 1
- Operating range: 100 m maximum
- Five-feet  of USB extension cable

The BlueTooth USB adapter adds BlueTooth technology to your existing USB PC or notebook to let it work with all devices enabled with BlueTooth v1.1 technology. It offers built-in security using 128-bit encryption and authentication, allowing safe wireless access with plug-and-play convenience. It supports Microsoft Windows 98 SE, Me, 2000, and XP.

## Ways To Talk To Your PC (Voice to Text)

Developers use the phrase "speech recognition" to refer to speaker-independent command and dictation vocabularies, reserving "voice recognition" for biometric or security applications of a person's voice.

### Dragon NaturallySpeaking 7

Recently ScanSoft introduced NaturallySpeaking 7, which brings speech recognition into the home and office mainstream. According to the company, the new release increases accuracy by 15 percent, enabling users to achieve accuracy levels of up to 99 percent while dictating up to 160 words per minute. NaturallySpeaking 7 is available in Standard ($100) and Preferred ($200) editions.

### IBM ViaVoice for Windows

ViaVoice has a reputation for being arguably a step behind NaturallySpeaking for general PC use, but arguably a step ahead for serious dictation or heavy-duty word processing. Release 10 introduced a new speech engine with improved background-noise adaptation and one-key control of dictation/command modes. IBM's ViaVoice for Windows Release 10 is available in $30 personal, $60 standard, $70 Advanced, and $190 Pro USB versions.

Both Dragon NaturallySpeaking, and ViaVoice include text-to-speech (TTS) capability, for users who want natural-sounding voices available.

### AnswerPad 1.25.6—Internal AI Agent

AnswerPad has the ability to speak aloud and will let you speak to it by using speech recognition. It can even take on a much more friendly appearance when told to use one of the many available Microsoft Agent characters. AnswerPad will memorize any statement that is entered into it (Figure 5.12). A user may then ask it questions in plain English and Answerpad will try to answer by processing the information within its memory.

To download AnswerPad, go to: http://www.agentland.com/Download/Intelligent_Agent/601.html.

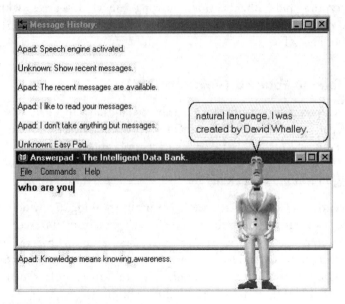

■ **Figure 5.12**  *Answerpad panel.*

## Assembled Basic MentorBot

The Basic MentorBot offers owners a usable robot under $1,000. It can navigate in a home without crashing into furniture, fixtures, or people. It can hold a conversation with humans and answer questions it has stored in its database. Figure 5.13 shows an assembled Basic MentorBot.

■ **Figure 5.13** *Assembled Basic MentorBot.*

The Basic MentorBot has two powered wheels and two swivel casters. It is slim enough to fit through a 24-inch doorway, but wide enough to stay upright on rough flooring. It gets its power from either eight D-cell batteries or a pair of gel-cell batteries, and its knowledge from an arsenal of cutting-edge sensors.

Whether the array microphone/speakers or BlueTooth headset/adapter are chosen, the Basic MentorBot will be able to converse. It will move about on command and respond to physical and verbal input from the user. The dual sonar ranging detectors prevent the robot from colliding with furniture, fixtures, or living beings. The wheel encoders and digital compass combine to measure direction and distance traveled, so the MentorBot knows where it is at all times.

The Basic MentorBot can also act as a companion and attendant for elderly. It will keep track of the elderly person, remind him or her to take pills or a meal and alert you when something is not right. It can also provide parents with an alternative to hiring a nanny or tutor for their children. It is able to teach simple math, distance, and physics and elicit a high degree of response from siblings.

# Enhanced MentorBot Kit

Based on Basic MentorBot Kit, the Enhanced MentorBot Kit gives users the complete companion robot.

## Digital Video Camera—USB

To give your MentorBot vision a WebCAM is the best choice. It should be top-quality CCD and provide 640 × 48-pixel resolution in color for video snapshots, navigational aid, face recognition, and remote video.

- CCD technology for color saturation and light sensitivity
- Color support, 24-bit (16.7 million colors)
- VGA sensor with software up to 640 × 480 pixels
- Digital video capture speed, 30 frames per second
- Adjustable-focus glass lens for crisp, sharp images
- Dual-function base for use on desktop PCs or notebooks
- Headset microphone for clear, low-noise conversations
- USB 1.1 interface for easy connection
- One 6-foot USB cable

■ **Figure 5.14**  *WebCAM and IR sensor.*

## IR Sensor Obstacle Avoidance and Data Input

The next step is to build in obstacle avoidance. The SRS Sharp IR detector is designed to detect IR light that is modulated (flashed on and off) at 40-kHz. When the detector sees a 40-kHz IR light source, it will signal this information to the computer. Instead of coding the 40-kHz cycle in the PIC, use a 555 timer gated to each IR LED.

The PIC cycles through each LED, turning it on and off a short period of time, about 100 times. The PIC monitors all the IR modules for hits and misses at the appropriate times. The PIC does this for each IR module, forms a code representing obstacles in the robot's path, and sends it to the MCU. The MCU then tells the PIC H-bridge controller what to do to avoid the obstacle.

## The Wireless MentorBot

The MentorBot should be able to follow you autonomously, while avoiding collisions with furniture and equipment. It should be able to maintain voice communication with you and an Internet connection while moving about. It should have a means of artificial intelligence (AI) and display a friendly face to you and yours.

A wireless configuration not only provides mobile freedom for the MentorBot, it places the robot in a network with other computers and Internet devices (Figure 5.15).

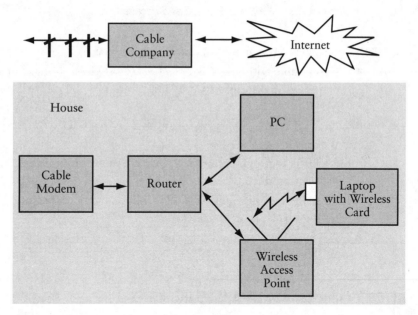

■ **Figure 5.15**  *MentorBot laptop in an Internet network.*

## Wireless 802.11b Communication

Take the power of an entire network with you—on your laptop computer. Surf the Web, retrieve data, and access other computers and peripherals with the speed and convenience of wireless networking. Network up to three computers (peer-to-peer/ad hoc mode) using a U.S.

Robotics 22-Mbps Wireless Access PC Card for each laptop computer and a U.S. Robotics 22 Mbps Wireless Access PCI Adapter for each desktop computer (Figure 5.16). For larger networks and for enhanced security features (infrastructure mode), add a U.S. Robotics 22 Mbps Wireless Access Point to your network design.

■ **Figure 5.16**   *802.11b PC card and access point.*

- Wi-Fi certified as interoperable at 11 Mbps
- Compatible with all 11-Mbps 802.11b wireless standard products; automatically adjusts to the fastest rate possible, 22 Mbps or 11 Mbps
- Network wirelessly at speeds up to 22 Mbps, twice as fast as most wireless networks
- Greater range and better area coverage than conventional 11-Mbps wireless products
- Ideal for getting multiple users online at the same time, sharing files, and accessing peripherals
- Advanced wireless security for protection and privacy; 256-bit WEP encryption

## ALICE Online Chatbot—A.I. Foundation

The ALICE A.I. Foundation is a nonprofit research and training organization devoted to the development and adoption of the Artificial Intelligence Markup Language (AIML).

The famous original free A.L.I.C.E. chat robot is still the most popular and most active AIML robot on the Web. The mother of all AIML robots, many other robot applications are cloned from the free A.L.I.C.E. brain. Download it from http://alice.sunlitsurf.com/downloads/.

## Advanced Software System

Advanced Software System (ASS) contains technologies for computer vision, autonomous navigation, human–robot interaction, and system architecture (Figure 5.17). ASS is used to control and receive feedback from the MentorBot. On the left side of the screen is an area to show you what the robot is seeing, a list of any recognized objects, any additional video input from an obstacle avoidance camera, and MentorBot movement controls.

■ **Figure 5.17**   *ASS control console.*

These conditions and actions can be used to create a wide variety of robot behaviors and to perform tasks. ASS allows up to 288 behaviors that can be run in any sequence or individually. Features include:

- Main camera; window displays what your MentorBot sees
- Personalized functions; create, save, and share functions
- Settings; change software settings and configure actions
- "If" conditions; set sight, sound, time, sequence, and sensors
- "Then" actions; specify actions to occur as a result of an event
- Behavior tabs allow multiple behaviors or link behaviors
- Object recognition; displays a list of recognizable things
- Train the MentorBot to recognize hundreds of objects
- Obstacle camera; shows you obstacles that the robot will avoid

- ASS dashboard; used to control the robot, see its x, y coordinates
- Sight; ASS can recognize colors, motion, objects, places
- Sound; give MentorBot a command: "be quiet" when dog barks
- Message; email the MentorBot to take a picture and email back
- Time; set an action to occur at 6 PM, or once a week
- Sequence; tell your MentorBot what to do next
- Move; train the MentorBot to follow you through the house
- Play a sound; play your favorite song, or read aloud
- Run a program; tell MentorBot to play the video it just captured
- Record a video and email it to you while you're away

More information can be downloaded at http://www.acrotek.com/products/MBkits.htm.

## Power System

You need to be able to turn things off at different times for testing. Here is where a power distribution board comes in. You can catch a glimpse of it in some of the pictures in the mobile base section. The board is nothing more than some filtering capacitors, switches and terminal blocks. You only need one mounted at this time but you should have a second one ready if needed.

### Batteries (Not Included) and Charger

The Basic MentorBot Kit makes provision for two battery selections. A pair of D-cell battery holders is included, and when loaded, they give two six-volt DC sources. If you plan to run your MentorBot a lot, heavy duty, rechargeable gel cells are more practical (Figure 5.18). Power-Sonic PS 6120s are recommended, and brackets are included.

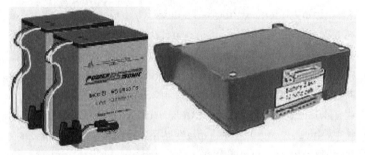

■ **Figure 5.18** *Gel cell batteries and charger.*

Two turn-on conditions are also defined: The first is the charging voltage that indicates that the charger is attached and that the power to the robot can be turned on without discharging the battery. The second is the minimum open-circuit voltage of the battery that indicates more than 10 minutes of energy is available to power the load.

This level of power (12 V DC, 12+AH) capacity weighs about 11 pounds which is 30 percent of the weight capacity of the platform. The motors and electronics draw about 3 amps when in use, which means 4 hours of use between charging. So for a robot work day from 8 AM to 6 PM (10 hours), the MentorBot would be available only 8 hours or so, with 2-hour charging intermissions every 4 hours.

For more information, go to: www.power-sonic.com/

## Body (Shell)

The simplest body shell to fabricate is a cylinder. Since many base units are round, this is a natural upward extension of that shape. Two slightly different diameter cylinders can be used for a variable height.

Consider making openings for access to the electronic, motors, and sensors. Make provisions on the lower exterior for a way of charging the batteries.

165

## Putting It All Together

Now that we've defined all the components, how do we put them together to make our own MentorBot? First of all, we separate the task into two parts—*above the shoulders and below the shoulders*.

We'll start with the *above-the-shoulders* portion which includes:

A. The laptop (your choice)
B. Digital video camera for color and face following
C. Wireless headset and adapter (BlueTooth)
D. Laptop BlueTooth access (BlueTooth PC Card)
E. Voice-to-text software (IBM Via Voice 10 Advanced)
F. Internal chatbot software (AnswerPad 1.25.6)
G. Text-to-voice software (PleaseRead Plus)
H. Wireless CF card (802.11b PC card)
I. Wireless access point (802.11b router)
J. Online chatbot (AI Foundation Silver ALICE chatbot)

Once the two PC card are plugged into the laptop, the software is installed, and the router is in your Internet access network (cable, DSL or satellite dish), you are ready to don the wireless headset and start talking with your laptop. Now, the *below-the- shoulders* robot to carry your laptop about should be to assembled.

The *below-the-shoulders* portion of your MentorBot includes:

K. Mobile robot platform with wheels, motors, and drives
L. Extra layers of the platform to raise your laptop to a useful level
M. Brackets to hold your laptop safely at the proper angle
N. Magnetic digital compass for direction and navigation
O. Sonar transducers and electronics for obstacle avoidance
P. Robot controller for robot movement, navigation, and safety
Q. Software to control movement, navigation, and safety
R. Batteries to power the robot and modules
S. Power distribution for getting power to individual modules
T. Voltage regulation to maintain the proper battery charge

In order to visualize where each module fits into your MentorBot, Figure 5.19 shows the letter identification itemized in the listings above. Follow how these modules are located on the MentorBot and relative to each other.

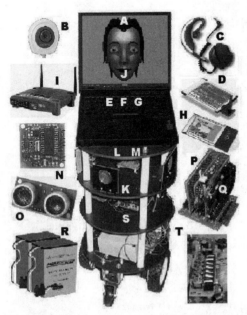

■ **Figure 5.19** *MentorBot assembly configuration.*

Once Steps K through T—the platform, extra layers, brackets, sonar, electronics, compass, software, power distribution, voltage regulators, and batteries—have been assembled, you are ready to operate your own "walking–talking" MentorBot. The first few hours will be spent in training the voice recognition, navigation, face recognition, and color following feature. After that, you need only enjoy.

## Assembled Enhanced MentorBot

The fully assembled MentorBot Kit is 70 cm (28 in) tall and 40 cm (16 in) in diameter. Because the ratio of height to width is less than two, the MentorBot platform is very stable. The height is close to desktop and is ideal for serving or to operate the laptop mounted on top.

The unused deck is available for the (Option A) version array microphone, speakers, and dispensing storage bins. The latter will be available as accessories.

■ **Figure 5.20**  *The Enhanced MentorBot.*

For more information and the latest pricing go to: http://www.acrotek .com/products/MBkits.htm

**Table 5.1** *MentorBot Kit Pricing*

| Item | Cost Per | Qty | Cost |
|---|---|---|---|
| Platform Kit | | 1 | |
| 16" base, casters, motors, drivers | $200 | 1 | $200 |
| HC controller and software | 190 | 1 | 190 |
| MC sonar and encoder driver | 50 | 1 | 50 |
| Poloroid ultrasonic ranger | 46 | 2 | 92 |
| Magnetic digital compass | 44 | 1 | 44 |
| Total: MB Platform Kit | | | 576 |
| Basic MentorBot Kit | | 1 | |
| MentorBot platform kit | $576 | 1 | $576 |
| MB basic software (MBBS) | 205 | 1 | 205 |
| Array microphone and speakers (A) | 150 | (1) | 150 |
| BlueTooth headset and adapter (B) | 150 | (1) | 150 |
| Additional two decks, and long risers | 28 | 2 | 56 |
| Manuals, cables, parts, and wires | 50 | 1 | 50 |
| Totals: Basic MB kit (A) or (B) | | | 987 |
| Enhanced MentorBot Kit | | | |
| Basic MentorBot Kit (A) | $987 | 1 | $987 |
| MB advanced software (MBAS) | 210 | 1 | 230 |
| Advanced speech software | 170 | 1 | 170 |
| 802.11b CD card and Ethernet access | 150 | 1 | 150 |
| Digital Web camera and IR sensors | 110 | 1 | 110 |
| Battery charger, locator and regulator | 100 | 1 | 100 |
| Body dhell with attachments | 90 | 1 | 90 |
| Manuals, cables, parts, and wires | 62 | 1 | 62 |
| Total: Advanced MB Kit | | | 1,879 |
| Laptop of your choice | ~$1,000 | 1 | ~$1,000 |

Prices subject to change—go to: http://www.acrotek.com/products/MBkits.htm.

## Remote Control and Video Display

To get full utilization from the Enhanced MentorBot, a remote control, video and voice remote are required. The HP iPAQ 5555 gives users BlueTooth and 802.11b WLAN access in the same package (Figure 5.21). The BlueTooth interface lets the iPAQ 5555 control action and converse with the MentorBot. With WLAN you can browse the Web, access email and view video from the MentorBot's onboard Web camera. The iPAQ 5555 also supports voice over IP (VoIP) when connected to a WLAN, and can use an optional headset to talk to MentorBot.

A small window under the iPAQ's navigation button is a built-in fingerprint scanner/identified. It also has the transflective, 64,000 pixel color

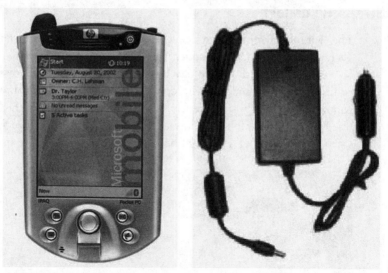

■ **Figure 5.21** *HP iPAQ h5555 PC and laptop charger.*

screen to view the same images being picked up by the MentorBot's Web camera.

### Battery to Laptop Converter

It is desirable to have the MentorBot battery charger also maintain the laptop computer battery charge. The best way to do this is with a 12-V DC-to-laptop converter. This product is a voltage-boosting 60 Watt DC/DC converter designed to power or charge a laptop from a 12-V DC robot battery. The 1060 series is in current production, and is available for retail or commercial sale in any quantity. For retail sale go to: http://www.powerstream.com/adc.htm.

### The Ultimate Approach

Humanoids may prove to be the ideal robot design for interacting with people. After all, humans tend to interact naturally with other humanlike entities; the interface is hardwired in our brains. The robots' bodies will allow them to seamlessly blend into environments already designed for humans. Undoubtedly, humanoids will change the way we interact with machines and will have an impact on how we interact with and understand each other.

## PINO the Humanoid

The Kitano Symbiotic Systems Project of the Japan Science and Technology Corporation under the Ministry of Education, Culture, Sports, Science, and Technology has been engaged in basic research on robust perception and behavior control using a humanoid robot. Such a robot is particularly suitable for this purpose due to its high degree of freedom and multiple sensor inputs. The PINO project (Figure 5.22) was started in November 1999 and became an open architecture in 2001.

■ **Figure 5.22** *PINO exterior.*

The key concepts for this project are:

- To develop a platform for perception and behavior research using multiple perception channels and high degrees of freedom
- To investigate robot design that will be well received by the general public
- To develop an affordable humanoid platform using off-the-shelf components and low-precision materials

The size of the robot is carefully designed to be that of an 18-month-old child (height 70 cm).

PINO's first public appearance was at The Venice Biennale in 2000. It was also in the Worksphere exhibition at the Museum of Modern Art

(MoMA) in New York. PINO was also featured in the music video "Can you keep a secret?" by Hikaru Utada.

## OpenPINO: Humanoid Robot Platform

OpenPINO (PINO-class humanoid robot platform) is an attempt to create a Linux-like open-source development community by disclosing technical information based on PINO. Software is disclosed under GPL (GNU General Public License) and electric circuits and mechanical design diagrams are released under GNU Free Documentation License. (Esthetic design is proprietary property and not a subject of open-source release.)

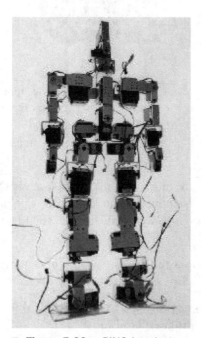

■ **Figure 5.23**  *PINO interior.*

OpenPINO platform (PHR-0001) as disclosed is a simple design, and has much room for improvement. It was only intended to be a minimum platform that serves as a starting point for collective effort.

Remember when Linux disclosed his initial Linux kernel, it was a collective effort of many interested people who contributed to the formation of the current Linux system. OpenPINO is the first attempt in robotics that tries to evolve through open-source movement. The hope is that this initiative will contribute to the promotion of scientific research and trigger faster growth in the robotics industry.

■ **Figure 5.24** *PINO circuit.*

The usual route to making a robot walk is to analyze the human gait and devise real-time control of the robot's joints to mimic it. At first, motors with a maximum torque of 25 kg-cm could not make Pino walk, but now, equipped with the genetic algorithm, it steps smartly using 7-kg-cm-torque motors.

In this algorithm, robots "learn" how to walk by themselves through trial and error. At the very beginning, Pino was just wiggling. But overnight, it had learned by itself to walk, when the evaluating parameter was properly set.

Yamazaki, Matsui, and Kitano started the Pino project in October 1999. Pino stood in April 2000 and started walking two months later.

ZMP INC. is selling PINO as a whole and as components only. Check http://www.zmp.co.jp/e_html/products.html for details.

If you are on a limited budget, go back to the new approach. PINO parts cost about $15,000 ($20,000 assembled).

**Table 5.2** *Ultimate Approach Pricing*

| Item | Cost/Item | Quantity | Cost |
| --- | --- | --- | --- |
| Open PINO parts (supplied by Zars) | $12,500 | 1 | 12,500 |
| MiniPC1-TX1 (main computer) | 750 | 1 | 750 |
| CCD camera (eye), microphone (ear) | 150 | 2 | 300 |
| MIT HandyBoard (control computer) | 250 | 1 | 250 |

*(continued on next page)*

**Table 5.1** *Pricing Summary (Ultimate Approach) (continued)*

| Item | Cost/Item | Quantity | Cost |
|------|-----------|----------|------|
| Linksys WRT54G Wireless-G Access | $220 | 1 | $220 |
| Jabra Freespeak BlueTooth headset | 180 | 1 | 180 |
| Application software package | 170 | 1 | 170 |
| Application software package | 160 | 1 | 160 |
| MS Windows XP Pro OS | 150 | 1 | 150 |
| Netgear MA701—Wireless CF card | 140 | 1 | 140 |
| IR sensor set | 130 | 1 | 130 |
| Application software | 120 | 1 | 120 |
| Battery: 12V 35 A/hr | 50 | 2 | 100 |
| Power distribution system | 80 | 1 | 80 |
| IBM via Voice Release 10 Advanced | 70 | 1 | 70 |
| Battery charger 2×12V | 60 | 1 | 60 |
| Digital compass | 50 | 1 | 50 |
| Power supply, Jameco 116089 | 40 | 1 | 40 |
| RS-232 DB9 to modular jack | 30 | 1 | 30 |
| Power supply, Jameco 116089 | 20 | 1 | 20 |
| Assorted cables and wires | 18 | 1 | 18 |
| Packet of small bolts/screws | 6 | 1 | 6 |
| 3-foot+ modular phone cord | 4 | 1 | 4 |
| Butt connectors, RS 64-3037 | 2 | 1 | 2 |
| Total | | | $15,550 |

**Contact:**

Yukiko Matsuoka
Mailto: open-pino@symbio.jst.go.jp
http://www.symbio.jst.go.jp/PINO/

## Summary (Build Your Own MentorBot)

Down the road, new innovations will provide a means to improve your MentorBot's performance. Components will be made faster, lighter, and stronger. Whole new concepts will allow you to revamp your MentorBot to give you added satisfaction and results. The changes may be in electronics, mechanics, and power sources. Mobile robots are making great strides today; the future will bring an accelerated pace of improvements.

## Intel to Make Robots Cheaper and Easier to Build

Intel announced Tuesday, March 25, 2003 that they will develop standardized robotics chips based on their own processors that will make it cheaper and easier for companies to produce robots. Intel's Robotics Engineering Task Force is devising reference designs for relatively small, inexpensive robots based around Intel silicon computer chips.

The company aims to standardize the internal electronics for robots, affording companies more time and resources to focus on developing such things as navigation systems, visual recognition systems, and artificial intelligence.

### Smart Machines

At the moment, most robots are guided by infrared or radio. In the future, researchers would like to use Bayesian navigation systems that employ artificial intelligence to let robots chart courses by matching information from a mounted camera to a memory map. This innovation would allow for robots that could automatically inspect industrial equipment or even take aerial photographs.

While larger companies are only now starting to look at the technology, two startup companies, Acroname and iRobot, are already using some of Intel's chips in their products.

"We're in discussions with Honda," says Jim Butler, director of the project at Intel Labs says. "We're talking to Samsung." Butler added that Intel is also trying to persuade the U.S. Defense Advance Research Projects Agency, with which it is already working, to take on primary responsibility for the project.

## Famous Battles

In 1988, a Carnegie Mellon chess program called Hitech beat former U.S. chess champion Arnold Denker, starting the chess wars. By the middle of the 1990s, artificial intelligence had become a common topic of household conversation. More chess battles between humans and computers—including Garry Kasparov versus IBM's Deep Thought and Anatoly Karpov versus Mephisto-Portrose.

Then, in May 1997, an IBM supercomputer called Deep Blue defeated Kasparov—at the time considered the greatest player in history—and the world's reaction was largely split between predictions of ascendant computers and scoffs at Deep Blue's brute-force approach to chess.

## Why Bother?

The Deep Blue-Kasparov battle wasn't the last big chess competition. In 2002, a computer called Deep Fritz took on champion Vladimir Kramnik in a match that ended in a draw.

With computers now regularly drawing with or beating chess champions—a feat long considered as a marker of intelligence—it appears machines have reached an important milestone. But with the years of research it took to get here, some might ask if it's worth the work.

If the goal of artificial intelligence research were simply to create artificial human minds, the answer might be no. It would make more sense for artificial intelligence researchers to spend time having and educating children.

## But This Is Not the Goal

Artificial intelligence research is as much about understanding our own intelligence as recreating it. How are you seeing these shapes, abstracting information and assessing its validity? Artificial intelligence work provides clues, with research into such things as pattern recognition, abstract thought and high-level deliberative reasoning.

As outlined above, artificial intelligence research has also led to valuable technologies. Many things we use each day owe it their existence: from search engines to expert systems that assist with disease diagnosis.

## Greater-than-human Intelligence

Perhaps the most significant goal is the creation of greater-than-human intelligence. Many problems in science and philosophy continue to stump human minds, and humans are beginning to create new problems for which we are ill equipped. Advanced nanotechnology, for example, would bring many benefits. But many people argue that preventing it from causing intentional or unintentional harm would require greater-than-human intelligence.

Is such intelligence possible? Most likely, yes, and probably quite rapidly after a machine reaches human-level intelligence. Several processes could contribute. Intelligent machines running on hardware that improves exponentially over time, for example, would experience a quantitative rise in intelligence along with that improvement. In addition, intelligent machines with access to their own source code could experience a qualitative rise in intelligence by essentially rewriting their minds.

175

In either case, intelligent machines would also become part of a feedback loop causing rapid development of greater and greater intelligence. In the former, they could be applied to the creation of better hardware that in turn makes them faster. In the latter, they could improve their ability to think about improving their ability to think.

## Preparing for the Future

Presently, few people are preparing for such a scenario. And yet it could prove to be one of the biggest challenges facing humanity in the next 50 years.

There is the question, for example, of what rights intelligent machines should have. Today, expert systems could make a more informed vote than most citizens in a democracy. Should they be allowed to do so?

And how will we interact with these beings? While possibly similar to us initially, they would share none of our biological heritage and their intelligence could rapidly become entirely different from ours. Battling Deep Thought, for example, Kasparov said that he sensed a new type of thinking that took him by surprise—not just faster thinking, but different thinking.

It is prudent to make every effort to ensure that different doesn't mean malevolent. To this end, some ethicists work alongside artificial intelligence researchers. At least one organization, the Singularity Institute for Artificial Intelligence, has the explicit mission of ensuring that future artificial beings will be friendly.

# 6

# Navigation Systems

## Sonar Navigation

Sonar sensing has some important properties that need to be carefully understood in order properly to exploit the sensing data. First, sonar transducers have a significant angular spread of energy known as *beamwidth*. In many systems, the beamwidth gives rise to large angular uncertainty in measurement. Some researchers have attempted to deal with this uncertainty by employing grid-based maps and repetitive measurements and relying on viewing targets from many locations. Localization with a grid map can be complex, so feature-based mapping schemes have become more commonplace after a means was presented for discriminating planes, corners, and edges using sonar data gathered at two positions. Later sonar sensors allow target discrimination at one position and target localization with high precision. Others have successfully demonstrated the localization capability of a mobile robot with a sonar array in a known environment using 3-D features.

The second important property of sonar systems is the appearance of phantom targets that are caused by multiple specular reflections. For example, a sonar sensor will see a virtual image of a corner due to the reflection from the wall in the outward and return paths in Figure 6.1. A credibility count is used to identify these phantom targets; however, this approach fails when the phantom target appears consistently from different positions, as is the case in the example of Figure 6.1.

The third important property of sonar systems is that, when sensing a planar wall, the sensor can only see the part of the wall that is orthog-

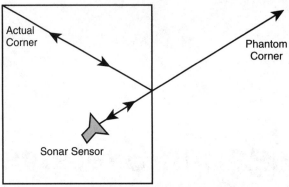

■ **Figure 6.1** *Phantom target example.*

onal to the line of sight—like phantom targets, this property results from specular reflection. Therefore, if the robot navigates along a wall, the robot sees the wall not as an entity but as a set of discrete, approximately collinear planar elements. Postulates must be made about the relationships between various sonar features during map matching. Furthermore, to reduce the risk of wrongly associating two features, the robot has to refrain from moving a long distance between successive scanning points during map building.

In successful mapping strategies:

■ All three types of primitive features recognizable by the advanced sonar sensor are processed to become part of a map. Discrete planar and corner elements gathered by the sonar sensor at various stages are merged incrementally to form *partial planes*. Planar elements are only merged to the adjacent partial planes to avoid falsely closing a gap, such as a doorway. Discrete edge elements do not partake in the process of forming partial planes, but they are still used to enhance localization accuracy and map integrity.

■ Not only does "plane to plane," "corner to corner," and "edge to edge" matching occur, as in other approaches, but the relational constraint between a corner and two intersecting planes is exploited to further improve the fidelity of map.

■ The partial planes are used to distinguish and subsequently eliminate phantom corner targets and edge targets caused by specular reflection.

■ Two implementations, based on two filters, are used to evaluate state transition equations, generate state-measurement cross covariance, and propagate error covariance matrices.

The map environmental model is presented and formulated as a statistical optimization problem that is solvable with a Kalman filter.

### Sonar Architecture

As shown in Figure 6.2, the communication backbone of the robot is an ISA AT bus with a 500-Mhz processor board controlling a custom sonar sensor card and a custom servo motion control card. The sensor control card sends transmit pulses and captures entire echoes from three receiving transducers. The motion control card provides PID control to the two DC drive motors. For every motor, an encoder is mounted on the actuation shaft (i.e., after a gear box) to generate feedback information that is not corrupted by backlash in the gearbox.

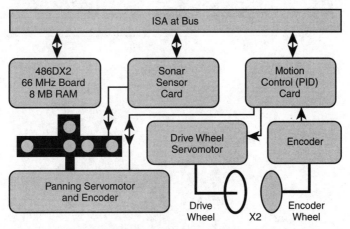

■ **Figure 6.2**   *Robot system architecture.*

### The Sonar Array

The sonar array illustrated in Figure 6.3 has a multiple transducer configuration that makes it possible to classify common indoor features into planes, 90-degree concave corners, and edges. At every scanning point, the sensor first simultaneously fires TR1 while scouting anticlockwise at 90 degrees per second to locate the directions of potential targets from the echoes on the three receivers. If classification is unsuccessful, the target is tagged as unknown, but range and bearing are still recorded to unknown objects.

### The Locomotion and Odometry System

The locomotion and odometry system shown in Figure 6.4 consists of drive wheels and separate encoder wheels that generate odometry measurements from optical shaft encoders. The encoder wheels are made with O-rings contacting the floor. These are as sharp-edged as possible to reduce wheelbase uncertainty and are independently mounted on linear bearings to allow vertical motion, and hence minimize problems of wheel distortion and slippage. This design greatly improves the reliability of odometry measurements.

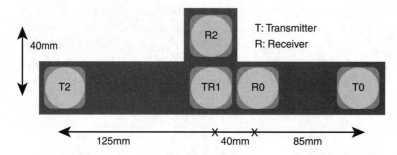

■ **Figure 6.3**   *Sonar array configuration.*

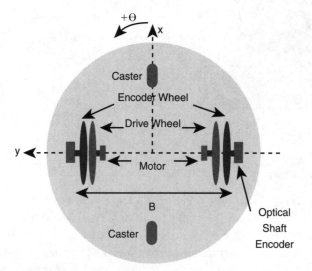

■ **Figure 6.4**   *Odometry system.*

### Growing Map Primitives

A snapshot of the map building scenario at stage $k+1$ is depicted in Figure 6.5. The robot has just moved to a new position and sensed a few new features. It is now ready to use some features for localization, and add the remaining features to the map.

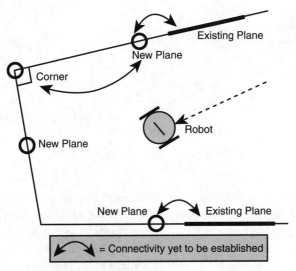

■ **Figure 6.5**  *Status of map and data fusion.*

Since the robot is operating indoors, discrete feature elements are assumed to come from a few planes; thus, they can be merged using a *collinearity constraint* to give a more realistic representation of the environment.

A planar measurement is fused to a partial plane if it satisfies the conditions depicted in Figure 6.5. A corner measurement is fused to an existing corner feature if it is close enough to it; otherwise, it is fused to two existing intersecting planes if it satisfies the conditions depicted in Figure 6.6. In a typical real environment, the artifacts on the walls, such as moldings, produce edges. Although excellent stationary landmarks for map building and localization, these cannot be considered as collinear with the nearby walls. Therefore, an edge is only fused to an existing edge if they are in the proximity of each other. For all grayed condition boxes in the figures, $\chi^2$ tests are applied. Every time a reobservation of a feature or relation occurs, the state of every map feature is updated because of their correlation. The unterminated endpoints of partial planes are projected to the new gradient determined by the new state parameters, whereas the terminated endpoints are recalculated from the intersections of all pairs of partial planes marked as terminated with each other.

### Distinguishing Phantom Targets

Local maps are preserved. Each feature in the local map has a parameter indicating which state it has been fused to. Therefore, the knowledge of where a particular map primitive was observed is available.

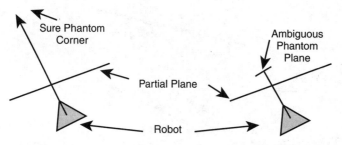

■ **Figure 6.6**  *Example of phantom targets treatment.*

When the map is sufficiently complete, some partial planes that block many phantom targets caused by specular reflection can be eliminated by checking the line of sight from the positions they were observed. If the phantom targets are too close to some partial plane they are considered ambiguous and are not eliminated. The strategies stem from the experimental observation that in a crammed indoor environment, such as a narrow corridor, corners and edges are more likely than planes to cause phantom targets. This is a result of the property that corners (and edges) return (some) acoustic energy in a direction opposite to its arrival, whereas planes reflect energy away from the arrival direction, except for normal incidence.

### Monte Carlo Localization (MCL)

Monte Carlo localization represents the probability density involved by maintaining a set of samples that are randomly drawn from it. By using a sampling-based representation, a localization method is obtained that can represent arbitrary distributions. It is faster, more accurate and less memory intensive than earlier grid-based methods (Figure 6.7).

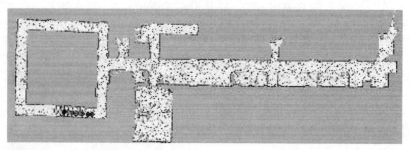

■ **Figure 6.7**  *Filtered sonar mapping.*

This particle-filter approach to mobile robot localization applies sample-based representations of the 3-D state space of the robot. MCL has been tested extensively using different types of sensors such as sonar, laser range finders, and cameras.

### Adaptive Real-Time Particle Filters

Real-time particle filters represent the belief resulting from mixtures of sample sets, thereby avoiding the loss of sensor data under limited computational resources. The size of the mixture is adapted to the uncertainty using KLD-sampling (Figure 6.8).

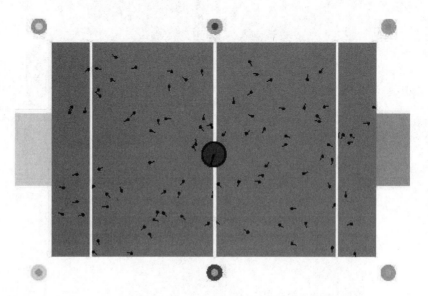

■ **Figure 6.8** *KLD filtered sonar particles.*

During localization, the sample set size is adapted by choosing the number of samples so that the sample-based approximation does not exceed a prespecified error bound. The animation shows the sample sets during localization using sonar sensors and laser range-finders. At each iteration, the robot is plotted at the position estimated from the most recent sample set. The blue lines indicate the sensor measurements. In this example, the number of samples was limited to 40,000. The time between updates of the animations is proportional to the time needed to update the sample set.

Note that the number can also increase if the robot becomes uncertain. Furthermore, the experiment shows that by simply using a more accurate sensor, such as a laser range finder, the approach automatically chooses less samples.

## Global Localization Using Sonar Sensors

MCL has the ability to represent the ambiguities occurring during global localization (Figure 6.9). The map generated during global localization uses the robot's ring of sonar sensors. Notice, that the robot is drawn at the estimated position, which is not the correct one at the beginning of the experiment.

■ **Figure 6.9**  *Global localized sonar.*

## Multirobot Localization

Figure 6.10 shows an experiment designed to demonstrate the potential benefits of collaborative multirobot localization. This approach makes use of the additional information available when localizing multiple robots by transferring information across different robotic platforms. When one robot detects another, the detection is used to synchronize the individual robots' beliefs, thereby reducing the uncertainty of both robots during localization.

The approach was evaluated using two Pioneer robots. Robot A performs global localization by moving from left to right in the lower corridor. Robot B detects Robot A as it moves through the corridor. As the detection takes place, a new sample set is created that represents Robot B's belief about Robot A's position. This sample set is transformed into a density tree, which represents the density needed to update Robot A's belief. The detection event suffices to uniquely determine Robot A's position.

■ **Figure 6.10** *Robots: A and B.*

### Recovery from Localization Failures

It is useful to add a small number of uniformly distributed random samples after each estimation step. These added samples are essential for localization if the robot loses track of its position.

Figure 6.11 demonstrates the ability of MCL to recover from localization failures. The data was recorded from a museum tour-guide MentorBot in the Smithsonian's National Museum of American History. To force MCL to lose track of the robot's position, severe errors were manually introduced into the odometer data: The robot was "teleported" at random points in time to other locations. These events can be detected by the sudden spread of the samples.

185

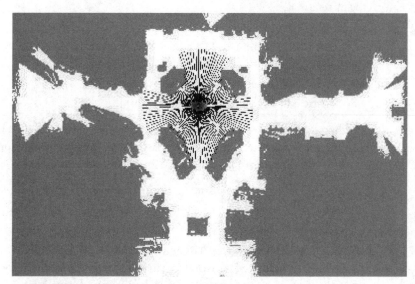

■ **Figure 6.11** *Failure recovery mapping.*

# Vision-Based Positioning

Vision-based positioning or localization uses the same basic principles as landmark-based and map-based positioning but relies on optical sensors rather than ultrasound, dead reckoning, and inertial sensors.

The most common optical sensors include laser-based range finders and photometric cameras using CCD arrays. However, due to the volume of information they provide, the extraction of visual features for positioning is far from straightforward. Many techniques have been suggested for localization using vision information, the main components of which are:

- Representations of the environment
- Sensing models
- Localization algorithms

The environment is perceived in the form of geometric information such as landmarks, object models, and maps in two or three dimensions. Localization then depends on the following two inter-related considerations:

- A vision sensor (or multiple vision sensors) should capture image features or regions that match the landmarks or maps.
- Landmarks, object models, and maps should provide necessary spatial information that is easy to be sensed.

The primary techniques employed in vision-based positioning are:

- Landmark-based positioning
- Model-based approaches
- Three-dimensional geometric model-based positioning
- Digital elevation map-based positioning
- Feature-based visual map building

Although it seems to be a good idea to combine vision-based techniques with methods using dead reckoning, inertial sensors, and ultrasonic and laser-based sensors, applications under realistic conditions are still scarce.

Clearly vision-based positioning is directly related to most computer vision methods, especially object recognition. So, as research in this area progresses, the results can be applied to vision-based positioning.

Real-world applications envisaged in most current research projects demand very detailed sensor information to provide the robot with good environment-interaction capabilities. Visual sensing can provide the robot with an incredible amount of information about its environment. Visual sensors are potentially the most powerful source of information among all the sensors used on robots to date. Hence, at present, it seems that high-resolution optical sensors hold the greatest promise for mobile robot positioning and navigation.

## Active Beacons

Navigation using active beacons has been with us for many centuries. Using the stars for navigation is one of the oldest examples of global-referenced navigation; technology has brought forward many other systems, such as lighthouses and, more recently, radio navigation.

For mobile robotics, laser, sonar, and radio (or microwave) are common media for navigational beacons. Because of this, most methods are "line-of-sight" dependent: there must be no obstructions between the mobile user and the beacons. Thus, these beacons have a limited range of use for a given number of beacons, and are mostly suited for local area navigation. However, they might well provide position fixes in a global frame of reference by transforming the user's position according to the relation between the local and global reference frames. The (longer wavelength) radio beacon systems tend to be more useful for global-scale navigation, due to the greater propagation distances available, and less suitable for indoor use, where multipath reflections from walls can introduce large inaccuracies.

Two principle methods are used to determine the user's position:

- *Triangulation* measures the bearing between the user's heading and a number of beacons.
- *Trilateration* uses a measurement of distance between a number of beacons and the user.

Almost all electronic navigation beacon systems used trilateration, as it is generally possible to measure time delays (and hence distances) more accurately than incident angles.

Most beacon systems can be subcategorized into one of the following transmission schemes:

- Scanning detectors with fixed active transmitting beacons
- Rotating emitters with fixed receiving beacons
- Scanning emitter/detectors with passive reflective beacons
- Scanning emitter/detectors with active transceiver beacons

## Radio Beacons

Global-scale radio beacon systems (e.g., Omega, Loran, GPS) tend to use scanning detectors with fixed transmitting beacons, because these allow for an unlimited number of users from a finite number of beacons. All transmitters are synchronized so that they transmit a continuous wave in phase. However, the receiver is not synchronized to this: the user can only measure differences in the time taken for signals to arrive from the various transmitters, not the absolute time. To calculate his position, the user finds the intersection of hyperbolic lines-of-position constructed using the difference in phase of signals received from two pairs of continuously broadcasting transmitters.

More localized beacon systems available commercially may use scanning detectors, transmitting beacons, emitters, or receiving beacons. The first two allow many users in one area; the first being more suitable for autonomous mobile robot control, as the position information is calculated at the mobile end. Emitter systems are more suited to tracking applications, such as motor cars around a racetrack. In the use of receiving beacons, round trip propagation delay time from user to beacon and back to user (or vice versa; generally in position monitoring rather than navigation situations) is measured—analogously to radar operation—to determine range. Using this exact range data, it is simple to calculate position by the intersection of circles around, ideally, at least three beacons. Possible user positions occur at the intersect of circles when the range to two and three transmitters is known.

## Ultrasonic Beacons

Ultrasonic systems generally use scanning detectors and transmitting beacons or emitters and receiving beacons.

Ultrasonics are frequently used in underwater situations, because sound has a much higher velocity (1500 ms$^{-1}$ c.f. 330 ms$^{-1}$) than in air. Also, it is possible to measure the incident angle of received signals much more accurately, allowing triangulation methods to be employed.

Ultrasonics are widely used for proximity detection. Occasionally, it is possible to combine the two, by introducing distinctive passive sonic beacons with unique reflection properties. By using trilateration against these beacons, a mobile robot can perform an absolute position fix, as well as find its position relative to any nonunique objects in the vicinity.

## Optical Beacons

Optical and laser systems often use emitter/detectors with passive retroreflective beacons, because these are very cheap. Laser energy is really the only form of transmission that can usefully be reflected without some form of intermediate amplification.

A great number of successful laser navigation systems have been demonstrated; an early example was the mobile Hilaîre robot, developed at the Laboratoire d'Automatique et d'Analyse des Systèmes, Toulouse, France. Hilaîre used groups retroreflective beacons, arranged in recognizable configurations, to minimize errors from reflections from other surfaces. Two rotating laser heads then scanned the area to determine the bearing to these beacons.

Other laser methods include a design using passive receiver beacons, from the Imperial College of Science and Technology, London. Here a rotating, vehicle-mounted laser beam creates a plane that intersects three fixed-location reference receivers. These then use an FM data link to relay the time of arrival of laser energy back to the mobile vehicle, so that it can determine distances to each beacon individually. A similar system is now commercially available through MTI Research, Inc., of Chelmsford, Massachusetts. This Computerized Opto-electrical Navigation and Control (CONAC) system is a relatively low-cost, high-precision positioning system, working at high speeds (25 Hz refresh), with an accuracy of a few centimeters.

# Environment Ranging Sensors

Most sensors used for the purpose of map building involve some kind of distance measurement. Three distinct approaches are used to measure range:

- Sensors based on measuring the time of flight (TOF) of a pulse of emitted energy traveling to a reflecting object, then echoing back to a receiver
- The phase-shift measurement (or phase-detection) ranging technique, involving continuous wave transmission as opposed to the short pulsed outputs used in TOF systems
- Sensors based on frequency-modulated (FM) radar; this is somewhat related to the amplitude-modulated phase-shift measurement technique.

## Time of Flight Range Sensors

The measured pulses used in TOF systems typically come from an ultrasonic, RF, or optical energy source. The parameters required to calculate range are simply the speed of sound in air or the speed of light. The measured time of flight is representative of traveling twice the separation distance and must therefore be halved to give the actual target range.

The advantages of TOF systems arise from the direct nature of their straight-line active sensing. The returned signal follows essentially the same path back to a receiver located in close proximity to the transmitter. The absolute range to an observed point is directly available as output, with no complicated analysis requirements.

Potential error sources for TOF systems include:

- **Variation in propagation speed.** This is particularly applicable to acoustically based systems, where the speed of sound is significantly influenced by temperature and humidity changes.
- **Detection uncertainties.** This involves determining the exact time of arrival of the reflected pulse. Errors are caused by the wide dynamic range in returned signal strength due to varying reflections of target surfaces. These differences in returned signal intensity influence the rise time of the detected pulse and, in the case of fixed-threshold detection, cause the more reflective targets to appear closer.
- **Timing considerations.** Due to the relatively slow speed of sound in air, compared to light, acoustically based systems make fewer timing precision demands than light-based systems and are less expensive as a result. TOF systems based on the speed of light require subnanosecond timing circuitry to measure distances with a resolution of about 30 cm (a resolution of 1 mm requires a timing precision of 3 picoseconds). This capability is very expensive to realize and may not be cost effective for most applications, particularly at close range, where high accuracies are required.
- **Surface interaction.** When light, sound, or radio waves strike an object, any detected echo represents only a small portion of the original signal. The remaining energy is scattered or absorbed depending on surface characteristics and the angle of incidence of the beam. If the transmission source approach angle exceeds a certain critical value, the reflected energy will be deflected outside the sensing envelope of the receiver. In cluttered environments, sound waves can reflect from (multiple) objects and can then be received by other sensors ("cross talk").

- **Ultrasonic TOF systems.** This is the most common technique employed on indoor mobile robots to date, primarily due to the ready availability of low-cost systems and their ease of interface. Over the past decade, much research has been conducted investigating applicability in areas such as world modeling, collision avoidance, position estimation, and motion detection. More recently, their effectiveness in exterior settings has been assessed. For example, BMW now incorporates four piezoceramic transducers on both front and rear bumpers in its Park Distance Control system.

## Phase-Shift Measurement

In phase-shift navigation systems, a beam of amplitude-modulated laser, RF, or acoustical energy is directed towards the target. A small portion of the wave (potentially up to six orders of magnitude less in amplitude) is reflected by the target's surface back to the detector along a direct path. The returned energy is compared to a simultaneously generated reference that has been split off from the original signal, and the relative phase shift between the two is measured, as illustrated in Figure 6.12.

$$\phi = \frac{4xd}{\lambda}$$

where

$\phi$ = phase shift
$\lambda$ = modulation wavelength
$d$ = distance to target

■ **Figure 6.12**   *Phase shift equations.*

For square-wave modulation at the relatively low frequencies of ultrasonic systems (20 to 200 kHz), the phase difference between incoming and outgoing waveforms can be measured with the simple linear circuit shown in Figure 6.13. The output of the exclusive-or gate (Figure 6.14) goes high whenever its inputs are at opposite logic levels, generating a voltage across the capacitor that is proportional to the phase-shift.

The advantages of continuous-wave systems over pulsed time of flight methods include the ability to measure the direction and velocity of a moving target in addition to its range (using the Doppler effect). Range accuracies of laser-based continuous-wave systems approach those of pulsed laser TOF methods. Only a slight advantage is gained over pulsed TOF range finding however, since the time-measurement problem is replaced by the need for sophisticated phase-measurement electronics.

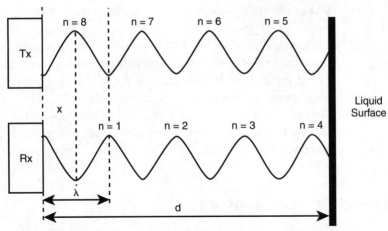

Relationship between outgoing and reflected waveforms, where x is the distance corresponding to the differential phase. (Adapted from Woodbury et al., 1993)

■ **Figure 6.13**   *Phase shift waves.*

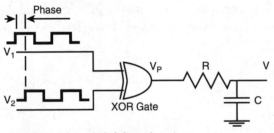

At low frequencies typical of ultrasonic systems, a simple phase-detection circuit based on an exclusive-or gate will generate an analog output voltage proportional to the phase difference seen by the inputs. (Adapted from Figueroa and Barbieri, 1991)

■ **Figure 6.14**   *Phase shift gate.*

# Landmark-Based Navigation

Landmarks are natural and instinctive visual cues to determine the whereabouts of a present location or a means to proceed to a particular location. Increased computational power and specialist hardware allows the real-time implementation of such ideas, with the ultimate goal of providing localization for agents such as mobile robots.

## Natural Landmarks

The main problem in natural landmark navigation is to detect and match characteristic features from sensory inputs. The sensor of choice for this task is computer vision. Most computer vision-based natural landmarks are long vertical edges, such as doors and wall junctions.

When range sensors are used for natural landmark navigation, distinct signatures, such as those of a corner or an edge, or of long straight walls, are good feature candidates. The proper selection of features also reduces the chances for ambiguity and increases positioning accuracy. A natural landmark positioning system has the following basic components:

- A sensor (usually vision) for detecting landmarks and contrasting them against their background
- A method for matching observed features with a map of known landmarks
- A method of computing location and localization errors from the matches

## Artificial Landmarks

Detection is much easier with artificial landmarks, which are designed for optimal contrast. In addition, the exact size and shape of artificial landmarks are known in advance. Many artificial landmark-positioning systems are based on computer vision. Some examples of typical landmarks are black rectangles with white dots in the corners or a sphere with horizontal and vertical calibration circles to achieve 3-D localization from a single image.

The accuracy of detection achieved depends on the accuracy with which the geometric parameters of the landmark images are extracted from the image plane, which in turn depends on the relative position and angle between the robot and the landmark.

A variety of landmarks is also used in conjunction with nonvision sensors such as bar-coded reflectors for laser scanners that are used in MDARS.

## Line Navigation

Line navigation has been widely used in industry. It can be thought of as a continuous landmark, although in most cases the sensor used in this system needs to be very close to the line, so that the range of the vehicle is limited to the immediate vicinity of the line. These techniques have been used for many years in industrial automation tasks, and vehicles using them are generally called automatic guided vehicles (AGVs). However, the techniques are not discussed in detail here since they do not allow the vehicle to move freely—the main feature that sets mobile MentorBots apart from AGVs.

The main implementations for line navigation are:

- Electromagnetic guidance
- Reflecting tape guidance or optical tape guidance
- Ferrite painted guidance, which uses ferrite magnet powder
- Thermal marker guidance

Artificial landmark detection methods are well developed and reliable. By contrast, natural landmark navigation is not sufficiently developed yet for reliable performance under a variety of and dynamic conditions.

The main characteristics of landmark-based navigation are:

- Natural landmarks require no modifications to the environment.
- Artificial landmarks are inexpensive and can have additional information encoded on them.
- The maximal distance between robot and landmark is significantly shorter than in active beacon systems.
- The positioning accuracy depends on the distance and angle between the robot and the landmark.
- Substantially more processing is necessary than in active beacon systems.
- Ambient conditions (such as lighting) can cause problems, such as landmarks not being recognized or other objects being mistaken for landmarks.
- Landmark navigation requires the robot to know its approximate starting location so that it knows where to look for landmarks. If this requirement is not met, very time-consuming searching processes must be employed.
- A database of landmarks and their location in the environment must be maintained.

## Map-Based Navigation

Map-based positioning (also known as "map matching") is a technique in which the robot uses its sensors to create a map of its local environment. This local map is then compared to the global map previously stored in memory. If a match is found, then the robot can compute its actual position and orientation in the environment. The pre-stored map can be a CAD model of the environment, or it can be constructed from prior sensor data.

The main advantages of map-based positioning are:

- It uses the naturally occurring structure of typical indoor environments to derive position information without modifying the environment.
- It can be used to generate an updated map of the environment. Environment maps are important for other mobile robot tasks, such as global path planning.
- It allows a robot to learn about a new environment and to improve positioning accuracy through exploration.

Disadvantages of map-based positioning arise because it requires that:

- There be enough stationary, easily distinguishable features that can be used for matching
- The sensor map be accurate enough (depending on the task at hand) to be useful
- A significant amount of sensing and processing power be available

## Map Building

Because the map building problem is closely related to the robot's sensing abilities, it could be defined as, "Given the robot's position and a set of measurements, what are the sensors seeing?"

Error and uncertainty analyses play an important role in accurate estimation and map building. It is vital to take explicit account of the uncertainties by, for example, modeling the errors by probability distributions. The representation used for the map should provide a way to incorporate newly sensed information into the map. It should also provide the necessary information for path planning and obstacle avoidance. The three main steps of sensor data processing for map building are:

- Feature extraction from raw sensor data
- Fusion of data from various sensor types
- Automatic generation of an abstract environment model

## Map Matching

Map matching is one of the most challenging aspects of map-based navigation. In general, matching is achieved by first extracting features, followed by the determination of the correct correspondence between image and model features. Work on map matching in the computer vision arena is often focused on the general problem of matching an image of arbitrary position and orientation relative to a model.

Matching algorithms can be classified as either *icon-based* or *feature-based*. The icon-based algorithm differs from the feature-based one in that it matches every range data point to the map rather than corresponding the range data into a small set of features to be matched to the map. The feature-based estimator, in general, is faster than the icon-based estimator and does not require a good initial heading estimate. The icon-based estimator can use fewer points than the feature-based estimator, can handle less-than-ideal environments, and is more accurate.

As with landmark-based navigation, it is advantageous to use an approximate position estimation based on odometry to generate an estimated visual scene (from the stored map) that would be "seen" by the robot. This generated scene is then compared to the one actually seen. This procedure dramatically reduces the time taken to find a match.

One problem with feature-based positioning systems is that the uncertainty about the robot's position grows if there are no suitable features that can be used to update the robot's position. The problem is particularly severe if the features are to be detected with ultrasonic sensors, which suffer from poor angular resolution.

## Geometric and Topological Maps

In map-based positioning, two common representations are used, geometric and topological maps. A geometric map represents objects according to their absolute geometric relationships. It can be a grid map or a more abstract map, such as a line or polygon map. The topological approach is based on recording the geometric relationships between the observed features rather than their absolute position with respect to an arbitrary coordinate frame of reference. Unlike geometric maps, topological maps can be built and maintained without any estimates for the position of the robot. As a result, this approach can be used to integrate large area maps, since all connections between nodes are relative, rather than absolute.

Map-based positioning is still in the research stage. Currently, this technique is limited to laboratory settings, and good results have been obtained only in well-structured environments. It is difficult to estimate how the performance of a laboratory robot scales up to a real-world application. The relevant characteristics of map-based navigation are that:

- They require a significant amount of processing and sensing capabilities
- Processing can be very intensive depending on the algorithms and resolution used
- An initial position estimate is required from odometry (or other source) to limit the initial search for features

The critical research areas are:

- Sensor selection and fusion
- Accurate and reliable algorithms for matching local maps to the stored map
- Good error models of sensors and robot motion

# Multirobot Communications and Coordination

A number of issues are introduced when more than one robot is present in a given locale concurrently. Without prior knowledge of each other's existence, they could well interfere with one another. However, if correctly programmed, and if the necessary communication network exists between them, they can cooperate in carrying out some task. Two advantages of cooperation will be considered here: improving speed and improving accuracy.

## Improving Speed

When using local area navigation, mobile robots are generally carrying out some set task (i.e., the navigation is the means, not an end). If a number of robots are cooperating in carrying out this task, then they will have to communicate positional data to each other to achieve this end.

When communicating positional information, a common reference should be used to compare positions. Thus, an absolute global or local positioning system should be used.

The communication link used between robots should ideally allow bidirectional transfers, with multiple access—allowing A to talk to B, without interference from C talking to D.

Given these conditions, considerable algorithmic advances can be made; these lie mostly in the higher-level "guidance" processing of the robot ("what should I do next?"), rather than the navigational side ("where am I?"). Thus, these improvements are mostly task dependent.

As an example, consider a number of robots searching a given area for some proximity-detectable object (e.g., using metal detectors). The main criteria dictating where a robot should look are:

- Have all of the objects already been found?
- Have all places already been searched?
- Am I within the area bounds?
- Have I already searched here?
- Has any other robot already searched here?

These questions can be answered by either continual polling of the other robots, or by each one carrying a complete current search state of the field, updated by events as they occur, or by querying a central search state database.

The method chosen aims to minimize both the amount of communication (and hence bandwidth) required, and the amount of redundant (potentially outdated) information in the system. The exact choice depends on the dynamics of the search.

## Improving Accuracy

While carrying out a navigational task, cooperating robots can provide augmentation for each other's navigational system, for better accuracy than the sum of their individual systems.

A good example of this is the *non-line-of-sight leader/follower* (NLOSLF) DGPS method. This involves a number of vehicles in a convoy that autonomously follow a lead vehicle driven by a human operator or otherwise. The technique employed is referred to as *intermittent stationary base differential GPS*, whereby the lead and final vehicle in the convoy alternate as fixed-reference DGPS base stations. As the convoy moves out from a known location, the final vehicle remains behind to provide differential corrections to the GPS receivers in the rest of the vehicles, via a radio data link. After traveling a predetermined distance in this fashion, the convoy is halted and the lead vehicle assumes the role of DGPS reference station, providing enhanced accuracy to the trailing vehicle as it catches up. During this stationary time, the lead vehicle can take advantage of on-site dwell to further improve the accuracy of its own fix. Once the last vehicle joins up with the rest, the base-station roles are reversed again, and the convoy resumes transit.

This ingenious technique allows DGPS accuracy to be achieved over large ranges, with minimal reliance on outside systems. The drawbacks to this approach include the need for intermittent stops, the reliance on obtaining an initial high-accuracy position fix (for the initial reference station), and accumulating ambiguity in actual location of the two reference stations.

## Scale of Navigation

The first stage in providing a navigation system for a mobile robot is to identify what scale of navigation is required.

Here, no requirement for global referenced position fixing is present, as the robot is only concerned with its position within the maze, and not with the absolute position of the maze on a larger scale.

On a local scale, the robot is concerned with its current position in the maze, and mapping all places visited in order to progress in solving the maze. At this level, the actual maze-solving processing is not considered; it is enough merely to produce the navigational and tracking information for the next level of processing to provide vehicle guidance information.

This local navigation has two distinct parts:

- Detection and categorization of walls and obstacles in the maze
- Determination of current position in maze

In mapping the maze, either walls or the path taken can be recorded. Depending on which of these is chosen, the requirements of the navigation system vary slightly.

If mapping walls, as well as sensing their position around the robot, the distance traveled by the robot must also be measured. If mapping the path traveled, this must be measured and recorded by some other means; the detection of walls is still necessary, however, to take a central path through the corridors.

The detection of walls can occur at either a local or personal level, depending on the technology used (proximity or contact). The path traveled can likewise be measured in a locale- or self-relative reference frame (local area position fix or dead-reckoning method).

## Viable Systems

Due to the large constraints of the robot specification, in terms of physical attributes and power consumption, not all applicable navigation systems can be used.

### Wall Detection

For wall detection the following techniques are available:

- (Ultra)sonic ranging
- Light-based ranging
- Tactile (contact) sensors
- Vision-based sensing
- Absolute position referenced onto a (previously recorded) map

Clearly, this last method is not possible, as having a pre-made map would nullify the aim of the robot. Vision also is of no use, as it would currently require too large (physically and electrically) an amount of processing for this application.

Of the other three, the technology for the application largely depends on the maze construction. The robot is circular—if *all* corridors are known to be of a width equal to the robot's diameter—then the tactile sensor is the simplest and most reliable method. If, however, the maze is constructed from larger corridors or rooms, this method might miss exploration of areas, and another method should be used.

Ultrasonic transceivers tend to be more reliable in detecting large surfaces (because light sensors suffer greatly from ambient light interference) and are available in quite small packages. Hence a number of these placed on the robot would provide a fairly reliable wall detection system. A contact sensor at the front of the robot might also be included for detection of collisions with objects invisible to the ultrasound.

### Position Determination

As discussed, knowing the position of walls about the robot is not sufficient for determining its position in the maze; some other distinguishing information is required. The walls themselves cannot provide this, because each one is not necessarily unique to the robot.

In determining position, the following could be used:

- An active beacon system working above the level of the walls
- Dead reckoning, giving a position relative to the starting location

The active beacon system is a viable option, as quite small, low-powered devices are available. Two considerable drawbacks exist, however. Most commercial systems are quite expensive, as they require a fair number (at least three) of beacons to be purchased, along with the processing power for their coordination in the environment. They are also quite intrusive, which reduces the autonomy of the robot. While this could be justified, a more self-contained solution would be preferred.

Dead reckoning, using odometry sensors, can provide good enough accuracy over short distances, and easily meets the physical constraints. It does, however, suffer from cumulative error, which requires periodic correction. Due to the nature of maze solving, a great deal of backtracking occurs; the corrections could be incorporated into this process by continuously comparing the actual position of walls sensed to their expected position according to the map made on the outward journey. By this, the accumulated error becomes a function of net displacement in the solution of the maze, rather than the total distance traveled throughout the mapping process.

## Two Contemporary Systems

The technology employed in mobile robot navigation is rapidly developing. Here, two relatively modern systems are studied, satellite based global positioning systems (GPS) and image-based vision positioning systems, which have the common feature of being under continual development. Between the two, many navigational requirements can be met.

### Global Positioning System (GPS)

In 1973, the American Defense Navigation Satellite System was formed as a joint service between the U.S. Navy and Air Force, along with other departments, including the Department of Transportation, with the aim of developing a highly precise satellite-based navigation system. In the 30 years since its conception, GPS has established itself firmly into many military and civilian uses across the world. Here, it is considered in the context of a device for navigating mobile robots.

When GPS was released by the U.S. Department of Defense (DoD), it superseded several other systems. It was designed to have limited accuracy available to nonmilitary (U.S.) users. Several methods of improving the performance have been developed to greatly increase the usefulness of the system for robots.

### GPS Segments

The *space segment* of GPS is twenty-four satellites (or Space Vehicles, SVs) in orbit about the planet at a height of approximately 20,200 km, such that generally at least four SVs are viewable from the surface of the earth at any time. This allows the instantaneous determination of user position to be determined, at any time, by measuring the time delay in a radio signal broadcast from each satellite and using this and the speed of propagation to calculate the distance to the satellite (the *pseudorange*). As a rule of thumb, one individual satellite needs to be received for each dimension of the user's position that needs to be calculated. This suggests three satellites are necessary for a position fix on the general user (for the x, y, and z dimensions of the receiver's position). However, the user rarely knows the exact time at which he is receiving—hence four satellite pseudoranges are required to calculate these four unknowns. The three segments of GPS are illustrated in Figure 6.15.

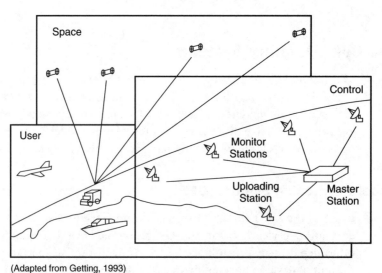

(Adapted from Getting, 1993)

■ **Figure 6.15** *The three segments of the GPS system.*

The satellite data is monitored and controlled by the GPS *ground segment*—stations positioned globally to ensure the correct operation of the system.

The *user segment* is the mobile user and his GPS reception equipment. These have advanced considerably in recent years, to allow the faster and more accurate processing of received data. They typically contain preamplification, an analog to digital converter, between five and twelve digital signal processor (DSP) channels (each one is able to track a separate satellite transmission), and a processor for navigational data. Other elements that might be incorporated are differential GPS receiver/processing capability, received phase information processing, and reception capability for the second (L2) GPS frequency.

### Uses of GPS

GPS provides an accuracy of 100 m (95 percent of the time) to *standard positioning service* (SPS) users, due to the selective availability (S/A) errors introduced intentionally by the US military, for defense reasons. This can be improved to about 15 m (95 percent) for authorized precision positioning service (PPS) users. The SPS accuracy is not good enough to be individually useful for mobile robot navigation. However, when augmented by the benefits of differential techniques, GPS does become a viable method for global reference navigation.

203

The DGPS system operates by having reference stations receive the satellite broadcast GPS signal at a known site and then transmit a correction according to the error in the received signal to mobile GPS users. As long as the mobile user is in the proximity of the stationary site, he will experience similar errors, and hence require similar corrections. Typical DGPS accuracy is around 4 to 6 m, with better performance seen as the distance between user and beacon site decreases.

DGPS provides the resolution necessary for most global scale navigation purposes, as well as often being useful at the local scale. A few restrictions are present in situations where it can be used however. The following problems can greatly reduce DGPS (or GPS) usability:

- Periodic signal blockage due to obstruction
- Multipath interference from large reflective surfaces in the vicinity
- As a result of both of these drawbacks, GPS will not work indoors.

In situations where the above are only a problem on occasion (e.g., a robot that operates outside as well as indoors), combining DGPS with other navigation technologies can prove very effective.

Another common marriage of technologies uses (D) GPS for global-level navigation, and other systems for precision local navigation. A good example of this is the U.K. Robotics Road Robot, a autonomous construction device. The Road Robot incorporates the Atlas navigation control unit, which initially finds its (global) course location using GPS, after which it uses laser trilateration to navigate (locally) while carrying out its task. This produced reliable autonomous operation in testing.

## GPS Receivers

When choosing positioning systems for mobile robots, a large number of variables must be considered, including:

- Size of mobile transceiver
- Power requirements of mobile transceiver
- Positioning accuracy
- Cost of mobile units
- Cost of total implementation
- Ability to process differential GPS data
- Inclusion of integrated data communications for other purposes
- Time to first position fix (from a "cold" start)
- Update rate with movement
- Standardization or availability of equipment
- Portability or time to set up

A great number of modern commercial GPS receivers and OEM development modules come with differential correction capability, of which almost all follow the RTCM-104 standard for interfacing with the DGPS receiver and network. This allows a GPS receiver to be used according to requirements, and DGPS correction signals can be used from any appropriate source by connecting the relevant DGPS receiver.

An example of a commercial DGPS receiver is the Communication System International CSI SBX-1. This is an "OEM module," designed to be integrated into another manufacturer's system. It is ideal for mobile robot construction. It is rated at less than 1 W at 5 VDC, and has a footprint of 10 cm$^2$. Coupled with a suitable GPS receiver (with typically somewhat high requirements; e.g., 10 W, 20 cm$^2$ footprint), this would provide a good ground for mobile position fixes.

# 7

# Advanced Technology

FOR MENTORBOTS TO ADVANCE, TECHNOLOGY NEEDS TO improve. Advances are needed in AI, speech recognition, robot vision, sensory-based controls, robot control systems, robot programming languages, and human–machine interfacing. These applications must expand into childcare and education, business and household assistance, communications, elderly caregiving, robot companionship, and knowledge storage and retrieval.

Since the first days of civilization, humanity has had to ponder the meaning of life. Centuries later and still no closer to an answer, humankind has seen society evolve into an existence where machines are a commonplace and necessary part of everyday life. As machines perform more duties that were traditionally done by beings of flesh and blood, the question arises: What does being human and alive mean?

## Artificial Intelligence (AI)

For many years, scientists have developed enough AI techniques to build a complete artificial human being. The biggest challenge now is to find useful, commercial applications for these AI techniques. In most real-world applications, it's not necessary that the application learn to do new things: it already has enough knowledge to carry out all instructions it's given. Most human workers also don't have to learn new things in their daily work. So AI learning techniques will only be useful to generate new smart applications. The learning algorithms don't need to be a part of generated applications. Applications only need to have the possibility to adjust to the work that has to be done.

Smart applications can be divided into two groups:

- New applications that could not be done without the intelligent use of computers:
  - □ Applications that need a lot of computing power, such as weather forecasting and making mathematical models by analyzing real world data (data mining)
  - □ Applications that can't be done by humans because they are too dangerous, such as fire extinguishing, work in heavily polluted areas, and space missions
  - □ Real world simulators and computer games (virtual reality)
- Applications that can replace human workers or that make the work of humans easier:
  - □ Robot applications
  - □ Automatic information processing, such as helpdesks, expert systems, OCR, speech recognition, computer vision, natural language processing, and industrial process controlling

## Learning Algorithms

Learning algorithms should only be used to generate new applications, so they will only be part of applications that are used to create new applications (e.g., statistical applications used to make models that can be used in new applications). Every smart application needs to decide which actions must be executed. It can only make the right decisions if it has a good model by which it is able to predict the real world, and these models are generated by the learning methods. There are three different kinds of automatic methods:

- **Extracting knowledge from human experts by interviewing.** This can directly result in a model (through fuzzy logic) or the acquired data can be processed by the next method.
- **Analyzing real world data.** Statistical or neural network methods can do this; these methods find a solution for most real-world models. But it's possible that a real-world system is too complex and these methods may not be able to find the right model for it. In these cases, an AI learning system could be useful; the system could find any solution if enough computing power were available.
- **Learning by experimenting in the real world.** Humans are good at this, and this is the system's primary design. Of course, this is only an option when it's possible for an application to experiment in the real world where it can't damage anything and errors are allowed.

When experimenting in the real world is not desirable, it's possible to simulate the real world in the computer. This alternative is only useful if it takes less time to program the simulation than program the model manually.

## Attentive-Regard Behavior

Attentive-regard behavior is active when a person has already established a good face-to-face interaction distance with the robot but remains silent. The goal of the behavior is visually to attend to the person and appear open to interaction. To accomplish this, the motor system holds a gaze on the person, ideally looking into the person's eyes if the eye detector can locate them. The robot watches the person intently and vocalizes occasionally. If the person does speak, this behavior reverts to the vocal-play behavior.

## Turn-Taking Behavior

The goal of vocal-play behavior is to carry out a protodialog with a person, relevant when the person is within face-to-face interaction distance and has spoken. To perform this task successfully, the vocal-play behavior must closely regulate turn taking with the human. This involves a close interaction with the perceptual system to perceive the relevant turn-taking cues from the person (i.e., that a person is present and whether or not speech is occurring) and with the motor system to send the relevant turn-taking cues back to the person. Four turn-taking phases occur: 1) relinquish speaking turn; 2) attend to human's speech; 3) reacquire speaking turn; and 4) deliver speech. Each state is recognized using distinct perceptual cues, and each phase involves making specific display requests of the motor system.

# Social Behaviors during Human–Robot Play

Socially situated learning is served by exploiting the types of interaction that arise between a nurturing caretaker and an immature learner. In this case, the learner is an anthropomorphic robotic platform. Its primary sensory inputs include vision, audition, and joint rotation. Its outputs include vocalizations, head and eye orientation, and facial expressions. The robot is designed to be an altruistic system, similar in spirit to a human infant. That is, the robot starts off in a rather helpless and primitive condition, and requires the help of a sophisticated and benevolent caretaker to learn and develop. The interaction with the caretaker robot is purely social, much like a mother interacting with her infant. The capabilities targeted for learning are those social and communication skills exhibited by human infants within the first year of life.

Today, much of the learning-based research in robotics is targeted at training a robot to learn a specific task, model, or representation. Often the researcher decides *a priori* what task the robot is to learn (such as navigating around an office environment) and then sets out to engineer the learning task accordingly. The learning task is completed once the robot can perform the task to a desired measure of success. However, because the learning algorithm is carefully tailored to a specific task, a new learning algorithm must be painstakingly designed if the robot is to learn a different task. The design of learning algorithms for robots is a labor-intensive process, and it is proving difficult to scale current techniques to more complex tasks in more complex environments.

In contrast, new work explores how to design a more open-ended learning system (Figure 7.1), heavily inspired by the theories, observations, and experimental results of child developmental psychology. The heart of this research is to decide how to design an integrated learning system so that the learner can bootstrap from previously acquired skills and cognitive structures to learn new, more diverse, and more sophisticated skills. Human infants are the prime exhibitors of the kinds of learning a system should emulate, having a developmental profile in which earlier skills and competencies are progressively modified, adapted, and built upon to produce more sophisticated, diverse, or new abilities.

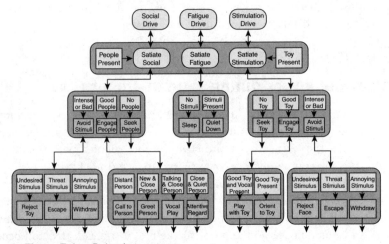

■ **Figure 7.1** *Behavior system.*

# Is AI a Threat for Our Future?

In the upcoming years, the amount of tasks that will be done by robotic AI will increase drastically. It may be possible that computer or robot can automate almost every task done by a human—the final result of the process that started with the Industrial Revolution. Is this a threat? With no need to work, and with unlimited resources available for all people, will the ultimate end of our capitalistic society be a socialistic society?

Some think that intelligent computers may take over our world and kick us out. This is very unlikely; a computer does only what it is told to do. Unless we give a computer a reason to develop dangerous behavior there will be no threat. Through evolution, man has developed many unpleasant characteristics like hate and selfishness, but because a computer is not part of natural evolution, it can only use evolutionary methods to reach the targets given it. A computer will never worry if it isn't able to reproduce itself or if it's going to be switched off.

Many people think it is very difficult to explain something like consciousness, because the human mind is not perfect and, unlike a computer, we can't view our own program code. The problem is that our mind gives us the impression "there is someone at home," because of the constant flow of thoughts in our mind. We like to label this flow of thoughts "I" or "me," But in reality this is just some states in our mind that are changing constantly (in a very clever way). This way of looking at the human mind is not new.

# Advanced Face and Head

Whereas anthropomorphic robots have bodies that look and physically act like the human body, anthropopathic robots are able to emote (Figure 7.2). The robots discussed in this section not only perceive and respond to human emotion, but are themselves possessed of an intrinsic emotional system that permeates their control architecture. For these humanoids, emotional state is not merely an outward expression, but can be used to influence the actions and behavior of the robot.

The robot Kismet is capable of using emotional modeling to guide its interaction with humans. Researchers at MIT used children and adults from different cultures to study how effectively Kismet can engage them through social interactions. Kismet responds not only to speech, but also to a variety of multimodal body language including body posture, the distance of the human from the robot, the movements of the human, and the tone, volume, and prosody of their speech. One of the underlying premises of the Kismet project is that emotion is necessary to guide productive learning and communication in general.

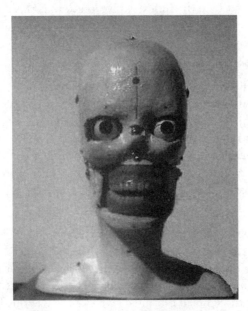

■ **Figure 7.2**  *Realistic face starts with a scull.*

To enact these behavioral rules, researchers designed small, mobile robots with six degrees of freedom on the face (two eyebrows, two eyelids, upper lip, and lower lip) and six degrees of freedom for the body (head yaw and pitch, neck, wheels, and back). Using these various degrees of freedom, the robots can exhibit compelling emotional responses computed using a periodic function generator that incorporates random variation as smoothly varying noise. In addition, the function generator also factors in the robot's current emotional state. The result is a group of robots that are inherently social in nature. The robots perform tasks based on their emotional "mood." For instance, if the robot is sad, it may perform actions slowly, whereas if it is angry, it may proceed violently.

Using twenty-four degrees of freedom, the robot expresses variations of seven emotional states including normal, happy, surprise, anger, disgust, fear, and sadness. Emotional state is based on the solutions of differential equations defined within in a three-dimensional coordinate space. For example, as the robot perceives a push, stroke, or hit from a human, it recognizes the action and maps it to an emotional space comprised of axes for pleasantness, certainty, and activation (sleep to arousal). As one might expect, the robot finds repeated abuse unpleasant, sleeps when it receives little stimulation, and uses recognition of objects to assign certainty. The goal is not merely to give the appearance of emotion. Rather, emotion is tied intrinsically to the way the robot performs tasks. The robot's interactions with the environment affect its emotional state and, vice versa, its emotional state affects the way it acts. Although its actions are derived from purely mathematical equations, the robot compels strong emotional responses from the human viewer.

## Advanced Arm and Hand Motion

The robot's arms should be loosely based on those of its human counterpart. Each arm should have six degrees of freedom, using a dedicated DC motor. Further, each arm should use two springs: a physical spring, supplemented by a virtual spring in software that adds the low-frequency characteristics. This combination allows the CPUs to simulate a passive spring-and-damper system (Figure 7.3).

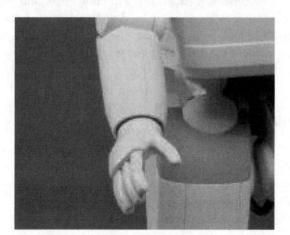

■ **Figure 7.3** *ASIMO arm and hand.*

The physical spring in the actuator doesn't closely emulate natural behavior because, like the electromechanical systems in an ordinary industrial robot, it's far too stiff. It does, however, implement a type of mechanical low-pass filtering that makes controlling the actuator easier. For example, the physical spring prevents chatter when an arm contacts a surface.

The software controller simulates a variable-compliance spring. The end result of this combination is that the overall springlike property gives each arm natural behavior; if an arm hits an obstacle or is otherwise disturbed, it deflects out of the way. The physical springs take up sudden loads, and the software springs take up "less-sudden" loads. More specifically, the arm compensates for a collision first using the physical springs and then by using the software springs.

Grasp planning for multiple finger manipulators has proven to be a very challenging problem. Traditional approaches rely on models for contact planning, which lead to computationally intractable solutions and often do not scale to three-dimensional objects or to an arbitrary number of contacts. An approach for closed-loop grasp control has been constructed that is provably correct for two and three contacts on regular, convex objects. This approach employs "n" asynchronous controllers to achieve grasp geometries from among an equivalence class of grasp solutions.

This approach generates a *grasp controller*—a closed-loop, differential response to tactile feedback—to remove wrench residuals in a grasp configuration. The equilibria establish necessary conditions for wrench closure on regular, convex objects, and identify good grasps, in general, for arbitrary objects. Sequences of grasp controllers and engaging sequences of contact resources can be used to optimize grasp performance and to produce manipulation *gaits*. The result is a unique, sensor-based grasp controller that does not require *a priori* object geometry.

## Advanced Bipedal Motion

Honda's new ASIMO Walking Technology (Figure 7.4) features a predicted movement control added to the earlier walking control technology to permit more flexible walking. As a result, ASIMO now walks more smoothly and more naturally.

### Predictive Movement Control

When human beings walk straight ahead and start to turn a corner, they shift their center of gravity toward the inside of the turn. With the new walking technology, ASIMO can predict its next movement in real time and shift its center of gravity in anticipation. The new "i-WALK" technology adds a sophisticated "predicted movement control" feature to the walking control technology of earlier Honda robots. As a result, ASIMO walks more smoothly, more flexibly, and more naturally, with instant response to sudden movements. Further, the range of movement of its

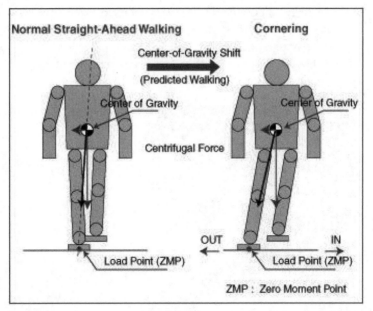

■ **Figure 7.4** *Honda's i-Walk technology.*

arms has been significantly increased, and the robot can now be operated by a new portable controller for improved ease of operation.

A human eases the impact of walking through a combination of structures and functions of movement. The former includes soft skin, ankles, and archlike structures comprised of several bones at the toe joints. The latter is ensured by the bending motions at joints when the plantae come into contact with the walking surface. Studies of human walking disclosed that the reaction from the surface tends to increase along with an increase in walking speed, even with the above-mentioned shock-easing functions. When walking at a speed of 2 to 4 km/h, the load to the leg/foot is 1.2 to 1.4 times greater than the body weight. At a speed of 8 km/h, the reaction load exceeds 1.8 times the weight. Although the robot must feature similar shock-absorption mechanisms, the structural measurements are not viable because they may reduce the robot's stability. Impact absorption is thus ensured through the precise control of each component. Based on the results of analysis, the specifications for the robot's legs and feet were determined.

Our sense of equilibrium is ensured by three mechanisms. Detection of acceleration is provided by the *statoliths*, three semicircular canals in the ear that detect angular velocity. The *bathyesthesia* of muscles and skin is responsible for detecting angles, angular velocity, muscular dynamism, pressures on plantae, and sense of contact. Also important is our visual sense, which supports and sometimes compensates for the sense of equilibrium. It also provides information required for normal walking. Thus, the robot system must incorporate G-force and six-axial force sensors to detect the conditions of legs/feet while walking, and an inclinometer and joint-angle sensors to detect the overall posture (Figure 7.5).

■ **Figure 7.5**  *ASIMO walks down stairs.*

Honda researchers found that the absence of toes had no significant effect on walking. More significantly, support is ensured by the base sections of the toes (i.e., balls of feet) and joint areas. Without foot joints, one cannot feel contact with the walking surface. Therefore, one is not only vulnerable to back-and-forth instability, but also less stable when crossing diagonally over an inclined surface. It is also impossible to ascend and descend stairs without knee joints. The absence of coxae makes autonomous walking extremely difficult. Examination results led to the selection of integrating coxae, knee joints, and articulation pedis in the humanoid robot.

## Realistic Movement

For a MentorBot to be effective, it must act naturally so that humans will feel at ease in its presence. To "act human" is a complex task requiring all the engineering technology available. The latest AI techniques, coupled with the new advances in mechanical motion, can make a MentorBot appear as a natural entity. The two most accomplished humanoids to date are Honda's ASIMO and Sony's SDR-4X. They each act very naturally.

Honda's ASIMO was designed to interact and service humans. Sony's SDR-4X was meant to entertain. Each has its place in the scheme of things, and the cost-to-performance ratios are similar. Time will tell which concept has the best acceptance and generates the largest market.

## Stereo Vision

One important technology that must be developed for any sort of free-roaming robot is stereoscopic vision. Otherwise the robot will not be able to accurately and quickly navigate toward its target, seeing and avoiding obstacles along the way. Real-time stereoscopic "machine vision" is one of the advanced technologies being developed and demonstrated for Urbie, a military reconnaissance robot.

In the past, robots used a single camera as the "eye," and human operators used the picture from that one camera to "see" obstacles and move the robot around to avoid them. This approach doesn't work well, though, when the robot is moving quickly or in a hazardous situation. There just isn't time for a human operator to analyze the image and react in time with commands to the robot. Also, for exploration of other planets, moons, asteroids, and comets, a robot must be autonomous. It takes minutes to hours for an image signal from a robot on a distant body to reach a human operator on Earth and for a command to be sent back to the robot. By that time, the robot could have fallen into a deep hole or gotten itself stuck.

Urbie's initial purpose is mobile military reconnaissance in city terrain. However, many of its features will also make it useful to police, emergency, and rescue personnel. The robot is rugged and well suited for hostile environments, and its autonomy will make Urbie ideal for working in dangerous situations. Such robots could investigate urban environments contaminated with radiation, biological warfare, or chemical spills. And, of course, such a robot will make an ideal space explorer.

The Defense Advanced Research Projects Association (DARPA) has enlisted JPL's Machine Vision Group in leading the design and implementation of this reconnaissance robot. Urbie is a joint effort of JPL, IS Robotics, the Robotics Institute of Carnegie Mellon University, Oak Ridge National Laboratory, and the University of Southern California Robotics Research Laboratory.

An outdoor mobile robot, such as the Navlab, needs not only information derived from appearance (e.g., road location in a color image, or terrain type), but also shape information (Figure 7.6). In some tasks, such as cross-country navigation, the 3-D geometry of the environment is the most important source of information. To build 3-D representations of the environment, an imaging laser range finder is used. 3-D vision for mobile robots has two objectives: object detection and terrain analysis. Obstacle detection allows the system to locally steer the vehicle on a safe path. Terrain analysis provides a more detailed description of the environment, which can be used for cross-country navigation or for object recognition.

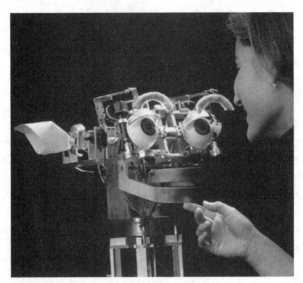

■ **Figure 7.6**   *Kismet stereo vision.*

Objects are detected from a range image by extracting the surface patches that are facing the vehicle. Neighboring patches are grouped into 3-D objects. The objects detected over many frames as the vehicle navigates can be combined into an object map. The resulting map can be used for navigating through the same region. Matching objects between observations is not very expensive in this case because only a few objects are needed to match in each frame and because a reasonable estimate of the displacement between frames from INS or dead-reckoning is assumed, so that the locations of the objects detected in one image can be easily predicted in the next image. The algorithm for building object maps includes provisions for removing spurious objects and for the optimal estimation of object locations (Figure 7.7).

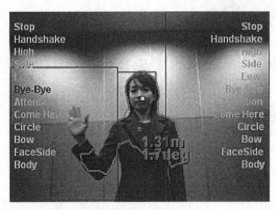

■ **Figure 7.7**  *Recognizing gestures.*

Object maps are not sufficient for detailed analysis. For greater accuracy, more careful terrain analysis is needed, along with combining sequences of images corresponding to overlapping parts of the environment into an extended terrain map. The terrain analysis algorithm first attempts to find groups of points that belong to the same surface, and then uses these groups as seeds for the region growing phase. Each group is expanded into a smooth connected surface patch. In addition, surface discontinuities are used to limit the region growing phase. This terrain representation is used in cross-country navigation systems.

As in the case of object descriptions, composite maps can be built from terrain descriptions. The basic problem is to match terrain features between successive images and to compute the transformation between features (Figure 7.8). In this case, the features are polygons that describe the terrain parameterized by their areas, the equation of the underlying surface, the center of the region, and the main directions of the region. If objects are detected, they are also used in the matching. Finally, if the vehicle is traveling on a road, the edges of the road can also be used for matching. As in the case of object matching, an initial estimate of the displacement between successive frames is used to predict the matching features. A search procedure is used to find the most consistent set of matches. Once a set of consistent matches is found, the transformation between frames is recomputed and the common features are merged.

217

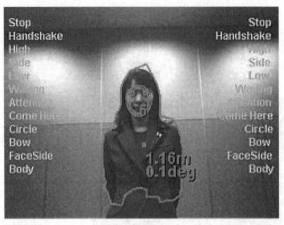

■ **Figure 7.8** *Recognizing faces.*

## Advanced DSP and Sensor Technology

To develop new techniques and methods in the area of environment perception, good knowledge of various techniques is required. It is essential not only to know the different sensor technologies, but also to have the skill to interpret signals by processing and analyzing measured data and know how to realize such systems in practice.

A new digital signal processing (DSP) technology forms a basis for the efficient implementation of environment perception systems. Furthermore, cost-efficient and reliable sensor technology is an essential requirement to design and build real-life robot applications. Intelligent sensors with advanced signal processing make it possible to build high-performance mobile robots without high costs.

## Ultrasonic Transducers

Capacitive ultrasonic transducers have been developed for mobile robot applications. Transducers are developed with the given resonant frequency, frequency bandwidth, and beam pattern. In addition, it is possible to predict transient response by applying mathematical models.

## New Ultrasonic Ranging System Based on Sensor Array

A prototype of a novel airborne ultrasonic ranging system for mobile robots allows a mobile robot to locate objects in a wide area surrounding it by obtaining both the distance of the object and the angle of incident. The obtained accuracy of the distance is about 1 cm and the resolution of angle is about 1 degree.

Another new mobile robot ranging system is based on a novel beam-forming method developed in VTT. This method takes advantage of the nonlinear filtering commonly used in image processing. Applied capacitive ultrasonic elements are also developed in VTT. New fast IC technology is used to carry out the computations of the proposed signal processing method.

## Advanced DSP-Architectures Applied to Sensor Signal Processing

The use of field-programmable logic devices, together with signal processors to implement enhanced environment perception systems, has been a major research area.

For example, FPGA-technology is applied to implement fast ultrasonic signal processing in mobile robot applications. FPGA is flexible technology for prototyping purposes and at the same time it has offered fast computation to realize various new DSP algorithms.

Autonomously moving vehicles need fast robot vision. High-speed DSP boards have been developed for special vision applications and pattern recognition that can be used in landmark detection and boom position measurement.

Simulation has been widely exploited in algorithm development, transducer design, and in the overall development of mobile robot control.

## Sense of Touch

Although robots with a sense of touch may be difficult to build, researchers are hoping for important pay-offs. *Haptic interfaces* allow users to "feel" objects that exist in a virtual environment. For example, if a user bumps into a tree or kicks a soccer ball within a computer-generated world, the joystick vibrates or provides force feedback to make the cyber-encounter feel real.

But quite apart from making better computer games, this technology could also help a surgeon practice a delicate operation without risk to a human patient or allow a geologist on earth to "feel" the texture of a boulder discovered by a robotic exploration device on Mars.

## Fingertip Touch

Scientists in Spain have developed a robotic finger with a sense of touch. It is made of a polymer that can feel the weight of what it's pushing and adjust the energy it uses accordingly (Figure 7.9). This is similar to the way humans use the sense of touch. If we pick up a delicate object, such as a flower, our fingertips sense its fragility and so grasp it lightly. We instinctively exert more force when holding or moving a heavier, more robust item because there is feedback between our sensations and muscles.

One way to make an artificial touch-sensitive limb, therefore, would be to equip it with delicate pressure sensors to provide this sort of feedback. This robotic finger is made from a "smart polymer," called polypyrrole. It expands in response to electric current and conducts differently in response to changes in pressure.

■ **Figure 7.9** *Finger tip touch.*

To convert the polymer's shape change into a bending motion, the researchers stick two thin polypyrrole films to either side of an insulating plastic tape. When they apply a positive charge to one of the films and a negative charge to the other, the first contracts and the second swells. This makes the whole sandwich bend.

These shape changes occur as electrons are pulled from or added to the chainlike polymer molecules in the films. This sets up a current of electrons from the negative to the positively charged film—so the bending motion uses up electrical energy.

An object in the path of the finger's moving tip as it bends gets pushed away at the same speed, regardless of its weight. For heavier objects, the finger simply pushes with more force and uses more energy. This adjustment happens because of the way the polymer film is squeezed against the obstacle. Pressure changes the packing of the polymer molecules, which alters the voltage needed to make it bend. In effect, the finger "feels" the resistance that its motion encounters.

## Sensate Skin

The human skin is the largest sensory organ of our body and of profound importance to how we interact with the world and with others. Yet, despite its significance in living systems, the sense of touch is conspicuously rare if not absent in robots. One research goal is to develop a synthetic skin capable of detecting pressure and location with acceptable resolution over the entire body, while still retaining the look and feel of its organic counterpart. Experiments are now being made with impregnating silicone materials (e.g., movie makeup) with conductive matter to make the silicone behave locally as a force sensitive resistor (FSR).

Giving the robot a sense of touch will be useful for detecting contact with objects, sensing unexpected collisions, and letting the robot know when it is touching its own body. Other important tactile attributes relate to affective content—whether it is pleasure from a hug or pain from someone grabbing the robot's arm too hard. In addition to developing sensate skin technology, pattern recognition algorithms to implement these perceptual abilities must be developed.

A new synthetic skin being developed is capable of detecting pressure and location with acceptable resolution over the entire body, while still retaining the look and feel of soft skin. A tactile sensing system has been developed in which strips of FSRs are placed in a grid pattern over the robot's core and under the silicone skin or fur. Using the homunculus distribution of sensing resolution as a guide, the density of sensors is varied, so that the robot has greater resolution in areas that are frequently in contact with objects or people. A distributed network of tiny processing elements is also being developed to lie underneath the skin to acquire and process the sensory signals.

## Sense of Smell

Although smell is the least significantly useful robotic sense, in certain situations, it can become important. One such situation is in identifying a location or situation the robot has experienced before. The second is in recognizing a dangerous situation where smoke or toxic fumes are present.

A completely new onboard "odor sensor" has been developed jointly by Kanazawa Institute of Technology, and New Cosmos Electric Co., Ltd. The developers believe that this is one of the first devices that can sense a particular odor with practical accuracy. Using this sensor, a robot will be able to detect a "burning scent," known to occur in the atmosphere preceding a fire.

## Software

Robot control systems have traditionally been built out of two distinct kinds of computational "stuff." On the one hand, there are the LISP-like symbolic systems from which theorem provers, planners, and expert systems are traditionally built. On the other hand, there are the parallel, reactive systems popular in behavior-based robotics and connectionist circles. The two classes of machines have complementary strengths and weaknesses. Our ultimate goal is to extend parallel-reactive systems to support higher-level cognitive tasks for which symbolic systems are typically used.

The target MentorBot is an expressive robotic creature with perceptual and motor modalities tailored to natural human communication channels. To facilitate a natural infant–caretaker interaction, the robot is equipped with visual, auditory, and proprioceptive sensory inputs. The motor outputs include vocalizations, facial expressions, and motor capabilities to adjust the gaze direction of the eyes and the orientation of the head. Note that these motor systems serve to steer the visual and auditory sensors to the source of the stimulus and can also be used to display communicative cues.

### The Low-Level Feature Extraction System

The low-level feature extraction system is responsible for processing the raw sensory information into quantities that have behavioral significance for the robot. The routines are designed to be cheap, fast, and just adequate. Of particular interest are those perceptual cues that infants seem to rely on. For instance, visual and auditory cues such as detecting eyes and the recognition of vocal affect are important for infants.

### The Attention System

The low-level visual percepts are sent to the attention system. The purpose of the attention system is to pick out low-level perceptual stimuli that are particularly salient or relevant at that time, and to direct the robot's attention and gaze toward them. This provides the robot with a locus of attention that it can use to organize its behavior. A perceptual stimulus may be salient for several reasons. It may capture the robot's attention because of its sudden appearance, or perhaps due to its sudden change. It may stand out because of its inherent saliency, as a red ball may stand out from the background. Or perhaps its quality has special behavioral significance for the robot, such as being a typical indication of danger.

### The Perceptual System

The low-level features corresponding to the target stimuli of the attention system are fed into the perceptual system. Here they are encapsulated into behaviorally relevant percepts. To elicit processes environmentally in these systems, each behavior and emotive response has an associated *releaser*. A releaser can be viewed as a collection of feature detectors that are minimally necessary to identify a particular object or event of behavioral significance. The function of the releasers is to ascertain if all environmental (perceptual) conditions are right for the response to become active.

## Fuel Cell Power Sources

In addition to the four main battery types mentioned in Chapter 3— sealed lead-acid, nickel-cadmium, nickel-metal-hydride, and lithium-ion—robot builders are working with fuel cells. Using a safe and economical fuel, the modern fuel-cell can provide more power at a lower weight than batteries do.

### Fuel Cells

A fuel cell works basically the same way that a battery does—by a reduction-oxidization reaction. However, rather than storing energy in the fuel cell itself, the cell is fed both fuel and oxidizer, which react in the cell to produce electrical power. The advantage is that the cell will run as long as you give it fuel; the disadvantage is that you need an external fuel source.

Fuel cells are currently used aboard the space shuttle and many other space missions: the models take in hydrogen and oxygen and produce electrical power and pure water at very high efficiencies. For more information, visit http://www.fuelcells.org.

A diagram of a fuel cell looks very much like a battery—you have an electrolyte, anode, and cathode. The fuel fed to a fuel cell can be a wide variety of stuff—hydrogen is used frequently since the output yields only power and water and is thus totally nonpolluting. Depending on the makeup, fuel cells can also run on most hydrocarbons such as methane or propane, although these are less efficient and produce carbon monoxide and dioxide as well. The fuel cell reaction is one of the most basic:

$$H_2 \rightarrow 2H^+ + 2e^-$$
$$H^+ + 1/2O_2 + 2e^- \rightarrow OH^-$$
$$H_2 + 1/2O_2 \rightarrow H^+ + OH^- \text{ (Overall E = 1.2V)}$$

■ **Figure 7.10** *A robot sized, 1-kw fuel cell.*

The biggest problem up until now has been the choice of fuels. Hydrogen is difficult to handle—it's a very light gas, can't be stored as a liquid for home use since it boils at about 20 K, and tends to turn metals that it is contact with brittle. You can store it in the same way that metal hydride batteries do, in a metal "sponge," but the metal weighs a lot, doesn't store a lot of hydrogen, and takes a long time to recharge. Other fuel cells use natural gas, although the technology behind the membranes needed to get good use is not as advanced.

A new fuel cell system that will run on ordinary gasoline and produce only carbon dioxide and water as exhaust is in development. This is a two-step process: the researchers have developed a "fuel processor" which converts ordinary gasoline into carbon dioxide and hydrogen gas. The hydrogen is then run through a standard fuel cell to generate power to run the robot. The technology behind this processor is still in the laboratory and it's going to take at least two years before a prototype is available.

## Sugar and Slugs Fuel Cells

Scientists have developed a bacterial fuel cell that runs entirely on sugar. The cell feeds a robot called Ecobot, which follows light around a room and must learn to collect its own energy supply. Ecobot's design is based on a similar robot "trained" to gather slugs to feed the biodigestor powering its energy system.

Scientists at the Intelligent Autonomous Systems Laboratory (IAS) at the University of West England have built a microbial fuel cell using *Escherichia coli* bacteria that feed off sugar. As the bacteria break down their food, electrons are produced and captured to power two motors. The motors move the robot towards light sources in burst motions.

Ecobot is being "trained" to balance its stored energy reserve against the requirements of its mission, replenishing its food and draining its waste. The process, known as *energetic autonomy*, was first tested in another robot, Slugbot, that hunted slugs to feed its off-board digestor unit. The digestor was designed to produce biogas for a methane fuel cell that generated electricity to power a battery pack. When "hungry," Slugbot could download its energy through the batteries. The team is also working on different types of fuels. Microbial fuel cells can be designed to run off garbage such as carrot peelings.

Slugs were chosen for the first project because they are easy to catch, are a major pest, and have no hard shell or skeleton. A robotic system could eventually be used in agricultural fields, with robots hunting the slugs at night when they are most active. Illuminating the field with red light would help the robots find the slugs, as the light would make the vegetation appear darker and the slugs brighter. A nearby fermentation station could power the robots' batteries from the slugs collected.

## Navigation

MentorBot navigation is a broad topic, covering a large spectrum of different technologies and applications. It draws on some ancient techniques, as well as some of the most advanced space science and engineering. Navigation techniques are explored in detail in Chapter 6.

## Summary

Doc Beardsley is an animatronics robot, a descendant of the mechanical humans and beasts that rang bells and performed other actions as parts of the clocks of medieval European cathedrals. Modern science, however, has carried Doc far beyond these ancient automata, endowing him with the ability to see, understand spoken words, and carry on a conversation.

Researchers at Carnegie Mellon University made the amusing, forgetful inventor as a literal embodiment of a computer interface. Doc performs for audiences, answering questions about himself. He claims to have been born on a mountaintop in Austria to a family of goatherds, and to have invented endless bread, the milk fed, the anti-snooze, and the foon (fork and spoon combined).

"In addition to paving the way for a future generation of theme park animatronics characters, the technology could lead to embodied personal digital assistants with personalities, interactive electronic pets, animated historical museum figures, and robotic waiters and salespeople," said Ron Weaver, a graduate student at Carnegie Mellon.

"Several layers of software drive Doc's apparent wit. Synthetic interview software, which includes speech recognition abilities, allows Doc to react to spoken questions. The technology, developed at Carnegie Mellon for use with video characters, gives a character a set of lines to deliver on given topics. This allows Doc to give appropriate answers to questions that match an anticipated query closely enough," said Weaver.

"If the question hasn't been anticipated, another layer of software takes over. A discussion engine tracks the questions and answers during a conversation and allows Doc to make relevant comments by keying off

226

individual words even if he doesn't understand a specific question," said Weaver. "And if that doesn't work, the discussion engine tosses the conversation back to the questioner."

"The discussion engine will first try to deliver a comment that is still relevant based on whatever individual keywords can be found in the text," said Weaver. "Failing that, the character gives a random comment that either pretends to reflect what is being discussed to try to keep [the conversation] going, or transfers the onus of the conversation back to the guest."

"One aspect of making Doc Beardsley a believable character is keeping the technology in a supporting role," according to Tim Eck, another Carnegie Mellon graduate student. "Character and story are the most important aspects to creating believable, entertaining characters." We are striving to provide the illusion of life, to create an entertaining experience, which is an important distinction. We are not trying to create artificially intelligent agents. We are creating the illusion of intelligence with time-tested show business techniques: drama, comedy, timing and the climactic story arc.

"The technology is not yet ready for the entertainment industry," said Eck. "The main reason [is] speech recognition technology. We believe once the overall accuracy of speaker-independent speech recognition is 80 percent or higher, applications such as ours will be seen in the entertainment industry. This will be approximately 5 to 8 years from now."

Another autonomous MentorBot, the CosmoBot, differs from standard rehabilitation tools in that it replicates and records a child's movements while providing monitoring access using the Internet. The robot operates in three modes: One mimics the gestures made by the child's physical motion or voice; another allows the child to record sound and movement and play it back, and the third leads the child through interactive games or stories that a therapist or teacher can download from a Web-based software interface, CosmoWeb.

Although one-on-one therapy is the most effective intervention for children with disabilities, there are difficulties. It is labor-intensive for therapists to keep a child's attention and continually motivate the child to stay on task. Also, children do not always perform necessary follow-up exercises or homework. Further, their access to therapy may be limited by financial constraints and the difficulties of arranging transportation to a therapist.

The need for repetitive actions and a stable environment provide a strong argument for using robotics in the rehabilitation process of autistic people, and children in particular. It is hoped that a robotic agent will disarm the child's fears and enable the child to relax, thus reducing the stress and pressure involved in the process of interaction. At the same time, a robot is able to adapt to the child and its environment in subtle ways and stretch the child's existing limits. A robot also has the dual advantages of being able to perform repetitive actions tirelessly and not appear to the child as a teacher figure which could be associated with learning. A robot provides the structure and repetition needed by autistic children to learn, while also being a friendly agent in their world. Children have become used to various forms of robots through television and toys in general and are able to relate to them in ways in which they cannot relate to other people. This allows a degree of trust to be built up, which in turn encourages interaction.

Since the overall goal is to encourage communication, the robot must provoke interest and enjoyment, or the child will not be motivated to interact. Once this goal is met, the robot must engage the child's attention without becoming threatening or dangerous in any way. Since the robotic platform used is sturdy and robust, it can be pushed around. It is also designed to avoid obstacles at all costs and moves relatively slowly; therefore, it will rarely come into contact with the children. The robot must perform two roles when interacting with the children:

- It must provide a stable environment in which the child feels secure. Autistic children generally prefer strict structure and limited environmental changes, and so repetitive actions must be used.
- It must also hold the child's attention and not become boring for the child. It must stretch the child's existing interactive and communicative abilities.

These two points mean that the development and control of the robot are of paramount importance. The robot must advance slightly ahead of the child, not becoming stale and also not becoming too advanced. This is complicated further by the individual nature of the autistic disorder.

Initial research and observations have directed the short-term goals of the CosmoBot project. It is apparent that it will be necessary to increase the scope of sensors available on the robot to allow a greater variety of input and, in turn, lead to more varied behaviors and responses. Also, to give the robot a longer "entertainment" span for the child, it will be necessary for it to adapt and provide a continually new experience without becoming totally unpredictable. This will entail the use of a short-term memory module and learning routine. Since the target group of autistic children may have many different strengths and weaknesses, it will also be necessary for the teacher or caregiver to be able to have a degree of control over the robot. A future development will involve the use of a teacher-moderated speech interface; however, since some autistic children are reluctant to use speech, this will be optional.

The individuality of autistic children requires the development of a robot that is flexible enough to meet the requirements of each child, and at the same time is structured enough to aim towards a similar goal—that of increased interaction—in different ways. A second important aspect relates to the method and result of behavior selection and stimulus–response pairing. The aim is to provide a response in an altogether different agent from the child, which is recognizable and coupled to a specific stimulation action. The decision mechanism must be robust enough to produce the same responses given a similar stimulus, while still being able to develop with the child's interactions and increase in complexity. This rate of complexity must also be rapid enough to maintain the child's interest while not developing too far ahead of the child.

Robots can make a valid contribution in the process of rehabilitation and potentially in the area of autism. They are able to produce consistent, repeatable, and reliable behaviors. This produces a stable environment for the user and builds a level of trust in the interactions present. At the same time, a robot is able subtly to adjust its actions to maintain the interest level without becoming greatly unpredictable. Robots have the potential to make a contribution in the area of autism rehabilitation, while also giving enjoyment to the children.

# 8

# Advanced MentorBots

## PaPeRo/R100

NEC's new robot PaPeRo (Figure 8.1) is small and light with many improvements over the first-generation prototype, Personal Robot R100 (Figure 8.1). PaPeRo's stand-alone architecture means it can easily be taken to many locations where two main types of research can be conducted: technology, such as improving its recognition and sensing ability; and its relationship between humans and itself. This fieldwork, it is hoped, will further the possibilities of introducing robots into our lives.

Using the latest in semiconductor and mechatronics technologies, PaPeRo's small (from 44 cm to 38 cm), light (from 7.9 kg to 5.0 kg) shape processes highly advanced reliability, safety, and communication capability functions.

The older R100 needed an external PC, which supplemented the processing power by processing speech and visual recognition remotely. PaPeRo can complete these tasks unassisted.

The number of recognized phrases (100 to 650), and words (300 to more than 3000) has increased considerably. Interaction is also more diverse due to PaPeRo's human affinity grades, which assign preference grades for each family, and memory and emotion technologies.

PaPeRo has a much more sophisticated and modularized architecture with a standardized interface for sensors and actuators that enables easy addition of new sensors, actions, and behaviors.

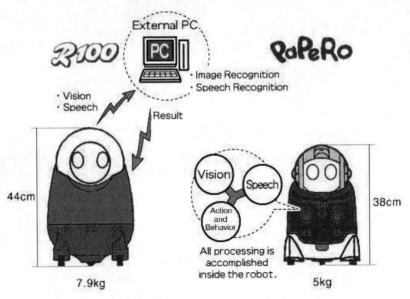

■ **Figure 8.1**  *R-100 and PaPeRo.*

Its actions and behaviors, such as dancing and conversation, can be programmed on an external PC with easy-to-use graphical editors. In this way, the robot and its software system have the potential for various research prototyping and development.

Details and specifications for PaPeRo include:

- Remote control signal transmitter transmits TV control infrared signal.
- Patting sense switch senses when it is patted on its head. When patted, it stops moving and presumes it has been scolded.
- Stroking sensor senses when it is stroked on the head. When stroked, it becomes happy and its preference for that person increases.
- CCD cameras as eyes, capture real-time images and recognize them. Using its visual capability, PePeRo walks smoothly, avoiding furniture and obstacles. It can find a human face, measure distance, and approach upon recognizing the person's face.
- LEDs are used for mouth and cheek for facial expression. LEDs on its ears indicate if it is listening and LEDs on its eyes indicate what it is looking at. These LEDs blink green when it is looking for people. They turn green when it finds a person and orange when it locates the person's face and attempts to track and recognize.

- Microphones for sound direction detection can estimate the direction of a voice using time difference detected by three microphones.
- Microphones for speech recognition on its head are used for speech recognition. As the robot always tries to track the face it is talking to, the direction of the microphone is kept facing to the person's mouth.
- Ultrasonic sensors sense distance to obstacles. When it walks, readings are used with visual data to supply supplementary information, especially when obstacles are in a blind zone or obstacles suddenly appear.
- Floor sensor watches for holes in front of the robot.
- Lift sensor underneath detects when the robot is lifted.
- Wireless modem terminal connected to a telephone line allows the robot to access the Internet. With its wireless transmission, actions and behaviors edited on the external PC can be immediately transmitted and executed by the robot.
- PaPeRo specifications
  - Height—385 mm
  - Width—248 mm
  - Depth—245 mm
  - Weight—5.0 kg
  - Battery duration—2–3 hours
  - Battery charge—2–3 hours
  - Number of recognized phrases—about 650
  - Number of speech phrases—about 3,000
- Output
  - Feet—two drive wheels (front), one free wheel (rear), max. speed: 20 cm/s
  - Head—up-down, left-right
  - Sound—two speakers
  - Face—eight LEDs in eyes, nine LEDs in mouth, two LEDs in cheeks, two LEDs in ears
- Other interfaces
  - Remote control signal transmitter
  - Video and audio output
  - Internet access

For more information, contact:

NEC Corporation
Tel: +81-44-856-2054
Fax: +81-44-856-2321
E-mail: robot@inc.cl.nec.co.jp
Web site: http://www.incx.nec.co.jp/robot/

## Robovie

ATR Intelligent Robotics and Communications Labs (IRC) has developed a robot called Robovie that has unique mechanisms designed for communication with humans (Figure 8.2). Robovie can generate humanlike behaviors by using its humanlike actuators and vision and audio sensors. In its development, software has been key. ATR-IRC has developed two important concepts in human–robot communication through research from the viewpoint of cognitive science: one is the importance of physical expressions using the body, and the other is the effectiveness of the robot's autonomy in robotic utterance recognition by humans. Based on these psychological experiments, ATR-IRC has developed a new architecture that generates episode chains in interactions with humans. The basic structure of the architecture is a network

■ **Figure 8.2**  *Robovie looking around.*

of situated modules. Each module consists of elemental behaviors to entertain humans and a behavior for communicating with humans.

In this project, the interaction between people and robots has been investigated, using the robot as an information media outlet. Fundamental issues of this research are how the robot interacts with people in town, how the robot can find its own purposes, and how such a system should be designed. The final goal is to build an information system for real-world robotic interaction.

## Current Status

Robovie was developed for both indoor and outdoor environments, as a test bed for the fundamental design of robots as information media. ATR-IRC achievements and future plans are:

- **Autonomous environment recognition of robots.** ATR-IRC has developed algorithms of state space construction by which a robot can make appropriate internal representation by itself with only sensor information.
- **A development method for intelligent robots.** ATR-IRC has investigated technologies to realize the robot's robust behavior in dynamic community environments. The robot architecture consists of situated modules that enable the progressive development of a robot that can achieve robust behaviors in indoor navigation.
- **Interaction between robots and people.** ATR-IRC investigated how information interchange can be achieved between people and robots. The robot system was developed with human–robot interaction in mind.
- **Social experiments by town robot.** To investigate the possibility of the robot working in a community, the robot system has been developed to work robustly and interact with people.

ATR-IRC has an experimental goal to build a robot that can navigate and work for hours in the area between the ATR-IRC building and a nearby restaurant.

## Future Direction

Robotics research began in the late 1960s; by the late 1980s, many autonomous robots were realized.

Robovie is the world's first communication robot that can balance itself on a two-wheeled pedestal. ATR has developed the Robovie series robots to achieve peer-to-peer communication between a human and a robot. Robovie-III has inherited the basic communication functions of the Robovie model developed in August 2000. These functions include a three-joint head, four-joint arms, skin sensors, mobility, omnidirectional image recognition, voice dialog capability, and ultrasonic distance sensor. The first feature of the new robot is its increased variety of communication motions in daily activities; with added waist, breast, and fingers, a greater variety of postures and motions can be expressed.

The second feature is the robot's hard-to-fall stability, achieved by adopting a moving mechanism using a coaxial two-wheeled inverted pendulum. An inverted pendulum is a balanced upside-down pendulum. Just as you may have balanced an upside-down broom on your palm in childhood play, a control system performs essentially the same operation. Robovie-III, with its two wheels coaxially mounted in symmetrical positions, can absorb changes in the upper body's center of gravity by controlling the torque and position of the wheels. Assisted by this mechanism, the robot can stand up or move while balancing itself on the two wheels attached to its sides against the dynamic movement of the upper body. This new feature enables the robot, for example, to talk about an object located under its knee, to talk with persons of different height by lowering its waist, to speak while looking back, to make humorous postures by moving its waist, or to point at a specific object with its finger. By combining these new motions with the previously developed dialog function and a variety of sensor information, the institute intends to conduct research on the communication principle between a human and an everyday communication robot, the method of expression used by a robot, and specific applications of the technology. This research has been performed under a contract with the Telecommunications Advancement Organization of Japan (TAO).

## Background of Development

The previous partner-type robot had multiple wheels or walked on two legs. However, from the viewpoint of communication, the two-legged robot presented safety problems when people came close to it. Also, with the added waist mechanism on a three-wheeled or four-wheeled pedestal, as used by previous Robovie models and other robots, there was the risk of losing balance and falling.

The previous research on Robovie's recognition function verified that a communication robot used in daily life needs to present motions similar to the natural movements of a human being to maintain communication with a person. For example, when a robot talks with a person about a nearby object, it needs not only to point at the object with its finger but also to move its neck (view line) as it first looks into the person's eye and then looks at the object. In real life, a person not only moves his neck but also uses his entire body. However, the swaying or bending motions of a robot body in everyday activities require the control of the robot's upper body, which has a large mass. One of the methods used to balance such movements is to attach a counterweight, but this increases the overall weight, thereby increasing the risk of danger when the robot falls down.

## Technology Highlights

Robovie-III, with the following added features, has solved the problem of increasing the variety of communication motions while avoiding the risk posed by the more varied motions:

- A mobile mechanism based on a coaxial two-wheeled inverted pendulum
- A parallel-linked motion mechanism between its breast and waist
- A waist unit with three degrees of freedom
- A finger-pointing mechanism

The coaxial two-wheeled inverted pendulum mechanism constantly maintains balance, so that any instability caused by the upper-body motion can be instantly corrected. This enables the upper body to move dynamically without losing balance. Also, the smaller number of wheels makes movement on uneven ground easier than with the previous Robovie.

The parallel link between the breast and waist increases the variety of communication motions that can be performed by the robot body. Specifically, with an added joint to the waist and breast, the robot can lean upward or bow with coordinated movements of the two parts. When the waist and breast were divided with a joint, the mechanism was simplified, with the center of gravity shifted lower to make it more difficult for the robot to fall. The motor is placed only in the waist, which connects with the breast through a parallel link. The lowered center of gravity reduces the positional shifting caused by the upper-body motions, thereby achieving dynamic yet safe motions of the upper body. A lowering motion of the waist makes the breast tilt forward, and a raising motion in the waist makes the breast stand upright, producing motions resembling those of a natural human body (Figure 8.3).

■ **Figure 8.3**  *Robovie socializing.*

A set of joints with three degrees of freedom is provided at the waist. The first joint is for twisting the waist to look back. This motion enables Robovie-III to look around without moving itself on the ground. The second joint is for tilting the upper body sideways by moving the waist to the left or right. The third joint is the parallel link for tilting the upper body downward or upward. Combinations of these waist motions can cancel the shifting of the center of gravity caused by breast motions, so the need for a counterbalance no longer exists. Combined motions can also keep the upper body upright while Robovie-III moves on uneven ground.

The last new function is finger pointing. The stick protruding from the arm tip can point to an object. This function allows the robot to point to a specific object among many. The previous Robovie model, with its rounded arm tip, was not very good at pointing.

Robovie is an everyday communication robot: that can independently interpret a scenario and perform daily activities by interacting with humans and the environment. The robot has its own activity rhythm (daily routine), which is the basis of its communications with humans or performance of duties. In previous research on communication robots, the goal had been a solution in a particular communication scene. Truly natural communication, however, involves a succession of scenes, and a solution for a single scene is insufficient. For this reason, the technique proposed for the everyday communication robot has more real-world applicability.

In contrast to a robot designed for factory production, a partner robot is designed for sharing daily life with a human. Currently available models include a pet robot, two-legged walking robot, toy robot, and communication robot. ATR intends to develop a partner-type robot that can maintain an equal relationship with a human beyond the master–slave relationship.

For more information, contact:

Hiroshi Ishiguro
ATR Intelligent Robotics and Communications Labs
Department of Adaptive Machine Systems
Osaka University, Suita, Osaka 565-0871, Japan
E-mail: ishiguro@ams.eng.osaka-u.ac.jp

## AMI

A Korean research team has developed a humanoid robot, bringing Korea hot on the heels of Japan in its drive to develop robot arm with intelligence and mobility. Humanoid robots use a high level of AI, thus technical difficulties have limited their development to a few advanced countries.

The latest humanoid robot (see Chapter 1) was the brainchild of a research team led by Hyun-Seung Yang, professor of electrical engineering and computer science at the Korea Advanced Institute of Science and Technology. "We are working on a humanoid robot with better intelligence with the ability to understand and express emotions," Yang said.

### Specifications

- Head
  - Eyes: 1 DoF × 2, neck: 2 DoF
  - Eyelid: 1 DoF × 2
  - Eyebrow: 1 DoF × 2
- Manipulator
  - Arm: 5 DoF × 2
  - Hand: 6 DoF × 2 (3 fingers each)

- Body
  - Waist: 1 DoF
  - Vehicle: 1 DoF × 2

The robot named AMI, meaning friend in French, is able to handle intricate manual tasks using its well-crafted hands and arms. It contains a number of high-powered sensors that constitute its visual, audio, and voice-mixing functions. It can listen to a human voice and recognize basic instructions while perceiving an object and measuring the distance involved in real time.

Its myriad sensor systems also allow the robot to detect and avoid obstacles while walking. A flat screen installed on the chest of the robot displays its internal operation levels as well as human emotions such as joy and sorrow.

At the heart of AMI's technology is the robot's self-contained power system, which makes it unnecessary for technicians to hook up the machine to a power connection for it to operate continuously.

Subprocessors are installed at key points, thus alleviating the pressure on the central microprocessor. The robot can be managed at a distance through a wireless LAN.

Yang said his research team spent the last two years designing AMI and its AI system and upgrading the prototype model, whose history traces back to 1991. Unlike its predecessor, AMI is the first attempt by the research team to create a humanlike robot for entertainment and other purposes in daily life, rather than for special operations (Figure 8.4).

A similar robot was developed by Honda Motor Co., which started humanoid robot research in 1986. The company focused on the "foot and leg-walking mobile function" that corresponds to the basics of human mobility, thus raising the stakes for the intelligent robot market in Asia.

Japanese developers have introduced a new generation of robots that are much smarter than their factory forefathers.

The research drive for robotics by KAIST, Korea's hub of high-tech engineering, is part of efforts to cut into the fast-growing market for robots designed for use in the home and office (Figure 8.5). For more information contact:

Yang, Hyun Seung, KAIST PhD Professor
AIM Lab, Dept. of EECS, KAIST, Korea
hsyang@cs.kaist.ac.kr +82-42-869-3527

240

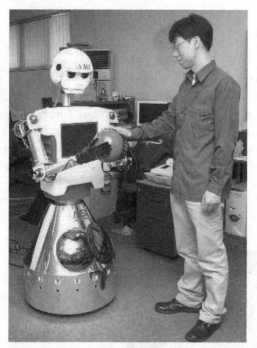

■ **Figure 8.4**  *AMI meets human.*

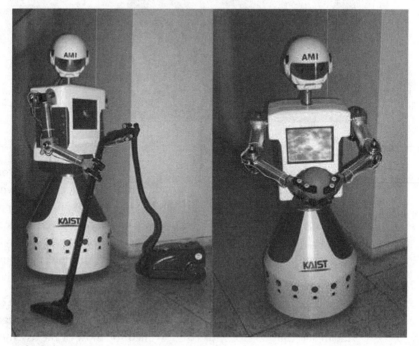

■ **Figure 8.5**  *AMI at work and play.*

## Robonaut

Robonaut (Figure 8.6) is a humanoid robot designed by the Robot Systems Technology Branch at NASA's Johnson Space Center in a collaborative effort with DARPA. The Robonaut project seeks to develop and demonstrate a robotic system that can function as an extravehicular activity (EVA) astronaut equivalent. Robonaut jumps generations ahead by eliminating the robotic scars (e.g., special robotic grapples and targets) and specialized robotic tools of traditional on-orbit robotics. However, it still keeps the human operator in the control loop through its telepresence control system. Robonaut is designed to be used for EVA tasks, that is, those which were not specifically designed for robots.

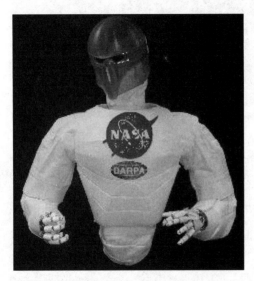

■ **Figure 8.6**　*NASA Robonaut.*

The challenge is to build machines that can help humans work and explore in space. Working side by side with humans, or going where the risks are too great for people, machines like Robonaut will expand the ability for construction and discovery. Central to that effort is a capability called dexterous manipulation, embodied by an ability to use one's hand to do work; the challenge has been to build machines with a dexterity that exceeds that of a suited astronaut.

A humanoid shape meets NASA's increasing requirements for space walks. Over the past five decades, space flight hardware has been designed for human servicing. Space walks are planned for most of the assembly missions for the International Space Station, and they are a key contingency for resolving on-orbit failures. Combined with a substantial investment in EVA tools, this accumulation of equipment requiring a humanoid shape and an assumed level of human performance presents a unique opportunity for a humanoid system.

While the depth and breadth of human performance is beyond the current state of the art in robotics, NASA targeted the reduced dexterity and performance of a suited astronaut as Robonaut's design goals specifically using the work envelope, ranges of motion, strength, and endurance capabilities of space-walking humans.

## Mechanism Design

The manipulator and dexterous hand have been developed with a substantial investment in mechatronics design (Figure 8.7). The arm structure has embedded avionics elements within each link, thus reducing cabling and noise contamination. Unlike some systems, Robonaut uses a chordate approach to data management, bringing all feedback to a central nervous system, where even low-level servo control is performed. This biologically inspired neurological approach is extended to left–right computational symmetry, sensor and power duality and kinematical redundancy, that enables learning and optimization in mechanical, electrical, and software forms.

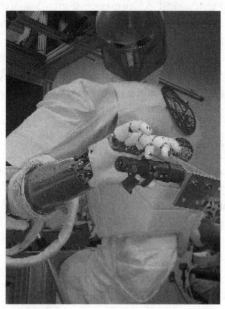

■ **Figure 8.7**  *Robonaut driving.*

The theory that manufacturing tools caused humans to evolve by requiring skills that could be naturally selected is applied to Robonaut's design as well. The set of EVA tools used by astronauts was the initial design consideration for the system, hence the development of Robonaut's dexterous five-fingered hand and human-scale arm that exceeds the range of motion of even unsuited astronauts. Packaging requirements for the entire system were derived from the geometry of EVA access corridors, such as pathways on the Space Station and airlocks built for humans.

## Sensors and Telepresence Control

Robonaut's broad mix of sensors includes thermal, position, tactile, force, and torque instrumentation, with over 150 sensors per arm. The control system for Robonaut includes an onboard, real-time CPU with miniature data acquisition and power management in a small, environmentally hardened body. Offboard guidance is delivered with human supervision using a telepresence control station with human tracking.

## Dexterous Manipulation

To meet the dexterous manipulation needs foreseen in future NASA missions, Robonaut is advancing the state of the art in anthropomorphic robotic systems, multiple-use tool handling end effectors, modular robotic systems components, and telepresence control systems. The project has adopted the design concept of an anthropomorphic robot the size of an astronaut in a space suit and configured with two arms, two five-fingered hands, a head, and a torso. Its dexterous arms enable dual-arm operations and its hands can interface directly with a wide range of interfaces without special tooling. Its anthropomorphic design enables intuitive telepresence control by a human operator.

### Fingertip Contact Sensors

A design for fingertip contact sensors for the second-generation tactile glove includes four sensors at the fingertip to provide a more accurate method of gathering contact location information. A mold was created in PRO-E to produce a silicon layer with force concentrators. This layer is bonded to a silicon sensor layer that can be shaped to provide the four small sensor areas.

### Tactile Glove

The Unit A left-hand tactile glove has less padding and fits more closely to the Robonaut hand than the original prototype. Including additional force concentrators on the fingers and thumb has also increased sensitivity. Software modifications to incorporate left-glove signals into the Robonaut control system are scheduled in the near future.

## Pro E Assembly Model

A new Pro/E assembly model of Robonaut in the free-flyer configuration (Unit B) features independently posable joints, subassemblies based on D-H (Denavit-Hartenberg) frames, reference data consistent with those used for controlling the robot, mass-related model parameters, and updated mass properties. It will be used in estimating the location of the robot's center of gravity, which depends on body posture. Interesting postures include lying down on an air-bearing sled, riding on the RMS in a stowed configuration, and leaning forward on a wheeled base. The current total body mass estimate for the system is 260 lbs, not including cabling, embedded electronics, and skin.

## FPGA Boards for Elbow and Shoulder

Motor controller boards using field programmable gate arrays (FPGAs) for the elbows and shoulders of Robonaut Unit B were reprogrammed with new FPGA firmware, and operation is significantly smoother with the updated firmware. Final tuning of the loop and feed-forward gains and determination of the best cutoff frequency for the stability margin compensation filter in the velocity loop remains to be completed. A separate computer has been dedicated to code development for the FPGAs and testing of the analog-to-digital (A/D) boards. This arrangement allows the development of new code to increase the performance of the FPGA boards.

## Unit B Analog System Checkout

The custom Robonaut Unit B Analog System has been checked out from the prototype remote sensor data acquisition board (MPA-ADC board) to the compact PCI Digital Signal Processor (cPCI DSP) interface board, including the serial interface cables, digital distribution board, and ribbon cables. The analog system has been tested to set gain, filter, and multiplex the signals from several types of sensors, including strain gauges.

A second cPCI DSP digital interface board has been completely assembled and is in the process of being checked out. It will be used to interface with the motor FPGA boards. Parts for a third board are being acquired. The third board will control the relay and brakes.

## Avionics Cluster Redesign

An alternative packaging solution for housing the avionics cluster has been proposed. In addition to other functions, the avionics cluster runs the tool roll axis in the Robonaut leg (TJ7) and any eventual foot DoFs. The design consists of a standard 4-inch roll joint (TJ7) nested into a shallow cavity in the mating 5-inch pitch joint (TJ6). The avionics are distributed in an annular region around the spacer connecting joints TJ6 and TJ7. It is a compact configuration with easily accessible but well-protected electronics and can be economically manufactured.

## Robonaut Experiment at Vanderbilt

A Robonaut experiment was conducted at Vanderbilt University as a follow-up to previous learning experiments. In the experiment, a teleoperator reached out and grasped a wrench five times each in nine locations. The experimenters at Vanderbilt correlated the motion to determine how Robonaut should grasp a wrench. The significance of the experiment was that the data collection was performed remotely using a virtual private network (VPN). During the experiment, the remote researchers were able to collect the data and view the motions of Robonaut by using the Robonaut graphical animation.

## Demonstrations

Robonaut demonstrations were supported for a group from the Naval Research Lab in Washington, D.C. and NASA management. The demos mark the first time Robonaut Unit A and Robonaut Unit B were run at the same time. Robonaut Unit B was run using the standard FITT-type telepresence gear, while Unit A was teleoperated using optical tracking. After the demonstration for the Naval Research Lab group, discussions were held for future collaborations in the autonomy area. For more information, contact:

NASA's Johnson Space Center
Robot Systems Technology Branch
1601 NASA Road 1, Houston, TX 77058
Phone: 281-244-2100 or
Email: schinfo

# RoboDog RS-1

The RoboScience RS-01 is a domestic quadruped robot. RS-01 has been designed from the outset to be fun to own. The development of a robot dog was a natural response to the proven demand for such products and also serves as a dramatic proof of concept for RoboScience's robotic technology. It is also entirely appropriate to the human environment.

The RS-01's brain is a miniature PC running the standard Windows operating system—a first for any commercial robot—allowing the RS-01 access to virtually limitless expansion and upgradeability. The RS-01 uses 802.11b wireless networking (Wi-Fi) to link to its owner's PC or home network, thus allowing it to be permanently online and giving it the versatility and functionality that consumers want from domestic robots.

RS-01 has a wide range of "senses" that allow it to interact safely with the world. Its senses allow it to balance, understand its location, and hear and see. Much as in biological creatures, onboard AI utilizes the information from these sensors to allow it to interact safely with the world.

The RS-01 has two operating modes:

- In autonomous mode, the RS-01 thinks and acts for itself without the need for human intervention. In this mode, it will have its own decision-making ability and is intended to perform useful tasks as well as just being fun, which makes it a significant step forward from current technology. Its large size and high performance will enable it to do things that other robots cannot, such as overcoming obstacles and navigating around a house or office. The primary user interface in autonomous mode is via voice recognition. The RS-01 will understand and act on up to sixty verbal instructions.
- In explorer or avatar mode, the owner can take control of the RoboDog via voice command or remotely from a PC. An onboard camera allows the owner to log into the robot via the Internet and drive it around its environment, seeing through its eyes, hearing through its ears, and so on.

RoboDog RS-O1 certainly walks the walk and barks the bark, and in most other ways—its hard, shiny carbon fiber coat and the absence of a wet nose excepted—gives its real life counterpart a close run for realism. RoboDog's specifications are:

- Movable parts: 16 degrees of freedom (DoF)
- Legs, 3 DoF each × 4

- Head, 3 DoF
- Tail, 1 DoF
- CPU, 32 bit, Pentium-style processor
- Storage, 64 Mb RAM—6 GB HDD
- Communication, wireless LAN
- Robot OS, Windows
- Host PC OS, Windows
- Motion control, 2 × RS MCU
- Sensors, color CCD camera
  - Miniature audio microphone
  - Accelerometer
  - Rangefinder
  - Navigation systems
  - Temperature sensor
- Audio output, two-way mono speaker system
- Battery, 26 V
- Run time, >90 mins in autonomous mode
- Charge time, approximately 70 mins
- Dimensions:
  - Height (standing) 685 mm (27 in)
  - Length, 820 mm (32 in)
  - Width, 400 mm (16 in)
- Construction, Composite monocoque construction
- Main materials, carbon-fiber, Kevlar, and magnesium
- Mass, approximately 12 kg (26 lbs)
- Color, metallic gold or natural carbon fiber black
- Max payload lift, 25 kg (55 lbs)

RS-01 breaks new ground in robot construction to achieve its unparalleled performance and size. Key technologies developed at RS include:

- The invention of new power train systems
- The first-time application in consumer robotics of lightweight carbon fiber composite monocoque construction
- State-of-the-art power electronics and battery technology

These breakthroughs are available to organizations interested in developing the next generation of robots as individual robot systems components (Figure 8.8).

■ **Figure 8.8** *Best friend RoboDog.*

For more information, contact:

RoboScience Ltd
Whittlebury, Northants
NN12 8TF England
Tel: +44 (0) 1753 44 8840
Email: info@roboscience.com

## Sony SDR-4X

Although still a prototype, SDR-4X has already reached marketability, because of its rich communication package and ability to adapt its movements to a variety of situations. Further development of its multiple sensor systems, performance control software, memory learning system, and a flexible biped walking mechanism may produce a biped-walking robot.

SDR-4X II is the enhanced version of SDR-4X, which was developed in 2002. The enhancements focus on its motion control and communication capabilities with people.

One of the most important issues, which must be addressed in biped walking robots, is the solution for fall-over. In SDR-4X II, in addition to its ability not to fall over, technologies have been applied to deal with the situation when it actually does fall over. When it is faced with a fall-over situation, it will adjust to make a motion control to avoid this in advance. If it cannot avoid fall-over, it will simultaneously take a posture to minimize the damage using its a fall-over avoidance motion control system. Here, actuators make compliant control to reduce the impact of ground contact in any direction. In the new model, the recovery motion control system has been enhanced for recovery from the fall-over situation.

The key to success for entertainment robots lies in their interaction with people. Safe design with respect to interaction with people is therefore vital. In SDR-4X II, a newly developed small-size actuator is introduced to enable high performance, such as active compliance control and active shock absorbance control. If something gets trapped in the moving parts of the robot, it is designed so that many of the touch sensors in the moving parts will detect the situation, leveling down the power of actuators to a safe level. Also, grabbing on the grip at the back of the robot and lifting it up can reduce the overall movement level and power down the torque at each joint.

SDR-4X II recognizes preregistered color identifiers and their configurations. An enhancement is the newly introduced map-building and identification capability, allowing the robot to self-recognize its estimated location in this learned configuration. From this capability, using color recognition, the robot is able to use its judgment to move to an adequate position or to make a move appropriate to its positioning or location.

With one added CPU for speech recognition and synthesis, SDR-4X II can internally process the recognition of large vocabularies (approximately 20,000 expressions), only previously possible by connecting to an external PC. In addition, the ability to memorize individual faces and names is improved by memorizing various words from spoken dialog with people, and matching spoken dialog to each individual. Furthermore, a preset topic scenario offers interesting spoken dialog, geared to each individual. The robot is capable of responding differently by utilizing the new phrase-driven dialog skill, thus enabling entertainment-rich conversation.

Including an original song called "Kiseki no Tabi," composed by Ryuichi Sakamoto with lyrics by Takashi Nakahata, more than 10 songs, 1,000 motions of various scales, and 200 scenarios of interactive spoken dialog have been developed as entertainment content for SDR-4X II. A design tool permits the design of autonomous behavior by combining these contents. Thus, it is possible to develop an entertainment robot that provides entertainment with natural autonomous behavior.

An internal battery charging system, as well as a separate station, is included in the design. The main specifications of SDR-4X II are:

- CPU—64 bit RISC processor (3)
- Main memory—64 MB DRAM (3)
- Operating system—Aperios (Sony's original real-time OS)
- Robot control architecture—OPEN-R
- Control program supplying media—memory stick
- Degrees of freedom (total: 38 DoF)
    - Neck: 4 DoF, fody: 2 DoF, arms: 5 DoF ($\times$ 2),
    - Legs: 6 DoF ($\times$ 2), independent hands (5 fingers)
- Sensors and switch
    - Distance sensor—Infrared distance sensor: head 1, hand 1 ($\times$ 2)
    - Acceleration sensor—trunk: X,Y,Z/3 axes, foot: X,Y/2 axes ($\times$ 2)
    - Angular rate sensor—trunk: X,Y,Z/3 axes
    - Foot sensor—foot 4 ($\times$ 2)
    - Thermo sensor—hand 1 ($\times$ 2), foot 1 ($\times$ 2), head 1, body 1, actuator 22
    - Touch sensor—head: 4, hand: 1 ($\times$ 2), shoulder: 1 ($\times$ 2)
    - Pinch detection sensor—overall: 18
    - Grip switch—back: 1
- Image input 110,000 color pixels CCD camera ($\times$ 2)
- Audio IN/OUT—microphone ($\times$ 7)/speaker ($\times$ 1)
- Input/Output
    - PC card slot (TypeII) ($\times$ 1)
    - Memory stick slot ($\times$ 1)
- LED display
    - Eye (4096 colors each), ears (1 color, 16 gradation), power (2 colors)
- Weight—approximately 7 kg (with battery and memory)
- Dimension (height $\times$ width $\times$ depth) approximately 580 $\times$ 270 $\times$ 190 mm

While improving the output performance of small actuators that drive each joint, a newly developed real-time integrated adaptive control system controls a total of thirty-eight joints in the robot's body in real time, based on information from various sensors gathered in real time. More advanced movements are realized by enabling the biped to walk on irregular and tiled surfaces; posture retention control is under external pressure. In addition, stable and flexible walk can be achieved by the real-time production of walking patterns, such as pace and rotation angle, in accordance with various situations.

Using two CCD color cameras for image recognition, the SDR-4X can detect the distance between itself and an object by processing the parallax of the two cameras. These cameras allow the robot to perceive an object and the range between itself and the object to automatically produce a route to make its way around the object.

In addition to image recognition, sound recognition, and sound synthesis technologies, communication and movement control technology based on memory is incorporated in the SDR-4X to further enrich communications with people. The SDR-4X can recognize an individual person by detecting the front facial image captured by a color camera. The robot can also detect the direction of a sound source and recognize a speaking individual by utilizing seven microphones located inside its head. By using embedded wireless LAN communication functions, the robot can synchronize data processing with an externally connected PC, which enables the continuous speech recognition of many vocabularies.

■ **Figure 8.9** *SDR-4X dancing.*

Information on people and the location of an object obtained through image recognition technology is used as short- and long-term memory information. Based on this information, the SDR-4X can achieve complicated communication and movements. Music and lyric data input into the robot cause it to produce a singing voice with vibrato; the composition of emotional, dynamic singing through voice synthesis can be realized to improve the robot's entertainment quality.

The SDR-4X was shown at the ROBODEX 2002, exhibition of "Robots as Partners;" SDR-4X II was shown at ROBODEX 2003.

## Prototype "SDR-4X" Key Characteristics

- Real-time integrated adaptive motion control and integrated adaptive control can be realized in real time for walking on irregular and/or tilted surfaces and retaining posture under external pressure (i.e., when pushed).
  - □ Ability to walk on an irregular surface up to 10 mm
  - □ Ability to walk up and down on a tilted surface, up to approximately 10 degrees
  - □ To prevent falling over under external pressure, the SDR-4X automatically performs footfall and step back to maintain its standing posture
  - □ If it falls over, damage is limited by the robot's flexible joint control and effective damage avoidance posture retention
- Real-time gait pattern generation control produces the necessary walking patterns by altering walking pace, cycle, and rotating angle in accordance with the situation and environment, based on information from the robot's sensors; this achieves a stable and smooth gait for autonomous walking.
- Real-world space perception technology, using two CCD cameras embedded in its head, permits the SDR-4X to detect the distance between itself and an object and perceive the range between. Based on this information, SDR-4X can automatically calculate a route to make its way around the object; seven microphones in the head make it possible for the SDR-4X to detect the direction of a sound source.
- Multimodal human interaction technology
  - □ Individual person detection, recognition learning technology
    - Ability to detect and recognize a face (front face) against a complex background
    - Ability to memorize up to ten individual faces through the learning functions
    - Ability to recognize individuals by the tone of their voice

253

- [ ] Continuous speech recognition and unknown vocabulary acquisition
  - Ability to recognize continuous speech for many vocabularies by synchronizing data processing with an externally connected PC via wireless LAN
  - Ability to learn and memorize new words not listed on its dictionary
- [ ] Conversation performance control technology based on its short-term and long-term memory. In addition to short-term memory functions to temporarily memorize individuals and objects, SDR-4X is equipped with long-term memory functions to memorize faces and names through more in-depth communications with people. Emotional information based on a communication experience will be memorized in long-term memory as well. By utilizing both short- and long-term memories, the SDR-4X achieves more complicated conversations and performances
- [ ] Speech synthesis and singing voice production
  - High level of entertainment produced by emotionally expressive speech and voice production synchronized with full-body performance
  - Capability to produce singing voice with vibrato through voice synthesis by inputting music and lyric data

- Safe design for interaction with people including a joint structure that does not trap hands and fingers in between joints

- Improved expression by adding more degrees of freedom; four axes in the robot's head and one axis in the wrist have been added to improve the robot's expression; five individually movable fingers are attached to each hand.

- Improved performance of small actuators. Compared to the small actuators used in prototype SDR-3X, a more advanced physical performance and response can be achieved by an approximate 30 percent improvement in the start-up torque, an approximate 15 percent improvement in rated torque, and an approximate 20 percent improvement in efficiency.

- SDR motion creator software system enables the easy production and editing of a variety of movements, such as dancing, using a PC. This software also includes automatic correction functions to prevent the robot's falling over while performing complex movements input by a creator or user.

For more information, contact:

Sony Corporation
3300 Zanker Rd., San Jose, CA 95134
(800) 222-7669
www.sel.sony.com

## Isamu

Kawada Industries, Inc. has developed a humanoid robot, HRP-1 Isamu, in conjunction with the University of Tokyo, Inaba Laboratory. Isamu is meant for use at home and at work.

Some of Isamu's specifications include:

- Height: 1.5 m
- Weight: 55 kg
- Degree of freedom, 32 degrees
- Speed up to 2 km/hr
- With two-camera stereo graphics input, the robot can recognize pre-entered human faces
- Grippers equipped with touch sensors
- Stair-climbing ability

This humanoid robot has technologies obtained from Kawada Industries' association with Tokyo University's H6 and H7 humanoid robots. The bipedal walk control system software was developed by the Inoue-Inaba Laboratory; the hardware and robotics structures, including servo level control system, was developed by Kawada Industries. Aircraft technologies were applied to the body frame, which led to a strong, light structure. A joystick can control the robot's variable walking pattern and speed of up to 2 km/hour. The robot can walk up and down 25-cm high steps. The hand grippers are equipped with touch sensors, and each hand can grip up objects weighing up to 2 kg (Figure 8.10).

In 2000, Kawada Industries joined a research and development project focused on the "transparent" humanoid robot development platform. This platform is based on the Humanoid Robotics Project (HRP) of the New Energy and Industrial Technology Development Organization.

255

■ **Figure 8.10** *Isamu waves.*

Kawada Industries plans to enhance the business of human-interactive motion control and uninhabited systems that have been developed in-house by electromechanical engineering technology obtained through helicopter research and development. Applications for human interactive motion control technology are mechanical systems for construction, disaster situation, handicapped aids, rehabilitation and training, and amusement. The newly developed Isamu will be used as a test bed for product development.

For more information contact:

Humanoid Robot and Intelligent Systems Devl Group
Aircraft and Mechanical Systems Division
Kawada Industries, Inc.
122-1 Haga-dai, Haga-machi, Haga-gun
Tochigi-ken 321-3325 Japan
Tel: 028-677-5737, Fax: 028-677-5707
Email: robocraft@kawada.co.jp

## Honda ASIMO

Although Honda is one of the world's biggest car companies, most people don't identify the Japanese auto giant with robots. But since the mid-1980s, Honda has been quietly developing a consumer-friendly robot in hopes that someday Honda Personal Robots will be just as ubiquitous as the Honda Accord.

Honda calls the robot ASIMO, which stands for Advanced Step in Innovative Mobility. ASIMO (Figure 8.11) is a humanoid robot, meaning he's designed to look and move as much like a human as possible. Robots are quite commonplace these days in the manufacturing world as more industries turn to robotic assembly lines to cut costs, but ASIMO was designed to be as much like a helpful, friendly humanoid as possible.

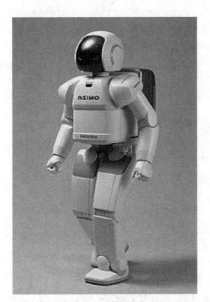

■ **Figure 8.11** *ASIMO walks.*

ASIMO is just under 4 feet tall, and weighs 115 lbs., Honda made ASIMO small because it knew that all the technological advancement in the world won't matter to consumers if ASIMO's appearance is even slightly aggressive or frightening. When you get to witness a humanoid robot up close and watch it move toward you—even one as small and cute as ASIMO—it takes a bit of getting used to after a lifetime of conditioning by scary robot movies. Honda purposefully designed ASMIO to be as cute and nonthreatening as possible, but also practical as well—a taller robot would have a harder time working at typical household chores, and ASIMO's low-to-the-ground body type makes him the perfect PC workstation operator.

ASIMO is not a toy robot dog that barks when you scratch his metallic belly—he's a multimillion dollar project that's far from completion. At this stage in his development, ASIMO can walk in a startlingly realistic, humanlike manner, he can speak preprogrammed sentences (but not engage in interactive speech with a human speaker yet), and he can perform simple tasks with his hands like wave, shake hands, and hand someone an object. But the real technological triumph is his walking, which took sixteen years to perfect to this stage. Give Honda another decade, and we'll be astounded at how closely ASIMO will be able to mimic our movements and behavior.

Honda is predictably oblique about its future plans for ASIMO. Currently, companies like IBM are able to lease the use of an ASIMO robot for corporate events and even simple tasks like greeting office visitors. But clearly Honda is setting the stage for a day in the not-so-distant future when consumers will have a Honda in the garage to take us where we want to go, and a Honda in the kitchen whipping up a gourmet dinner with one hand, burping the baby with another, and playing James Brown's "Live at the Apollo" on his internal audio system.

The key features of the new ASIMO include:

- Advanced communication ability thanks to recognition technology
  - Recognition of moving objects
  - Posture/gesture recognition
  - Environment recognition
  - Sound recognition
  - Face recognition
- Network integration
  - Integration with user's network system
  - Internet connectivity
- Recognition of moving objects using visual information captured by head-mounted camera
  - Detect the movements of multiple objects, assessing distance and direction
  - Follow the movements of people with its camera
  - Follow a person
  - Greet a person when he or she approaches
- Recognition of postures and gestures based on visual information
  - Can interpret the positioning and movement of a hand, recognizing postures and gestures
  - Can react not only to voice commands, but also to the natural movements of human beings
  - Recognize an indicated posture and move to that location
  - Shake a person's hand when a handshake is offered (posture recognition)

- □ Respond to a wave by waving back (gesture recognition)
- Environment recognition; ASIMO is able to assess its immediate environment, recognize the position of obstacles, and avoid them
  - □ Stop and start to avoid a human being or other moving object that suddenly appears in its path
  - □ Recognize immobile objects in its path and move around them
- Distinguish sounds; ASIMO can distinguish between voices and other sounds
  - □ Recognize when its name is called, and turn to face the sound
  - □ Look at the face of the person speaking, and respond
  - □ Recognize sudden, unusual sounds, such as that of a falling object or a collision, and face in that direction
- Face recognition
  - □ Recognize the faces of people who have been preregistered, addressing them by name, communicating messages to them, and guiding them
  - □ Recognize approximately ten different people
- Integration with user's network system
  - □ Execute functions appropriately based on the user's customer data
  - □ Greet visitors, inform personnel of the visitor's arrival by transmitting messages and pictures of the visitor's face
  - □ Guide visitors to a predetermined location
- Internet connectivity so that ASIMO can become a provider of news and weather updates

## Movement

When ASIMO's shoulder joint mounting angle was raised by 20 degrees, elbow height was increased to 15 degrees over horizontal, allowing a wider range of work capability.

Also, ASIMO's range of vertical arm movement has been increased to 105 degrees, compared to P3's 90-degree range.

The introduction of intelligent, real-time, flex-walking i-WALK technology allows ASIMO to walk continuously while changing directions, and gives the robot even greater stability in response to sudden movements.

The new ASIMO Walking Technology features a predicted movement control added to the earlier walking control technology. This new two-legged walking technology permits more flexible walking.

As a result, ASIMO now walks more smoothly and more naturally. When human beings walk straight ahead and start to turn a corner, they shift their center of gravity toward the inside of the turn.

Thanks to new ASIMO Walking Technology, ASIMO can predict its next movement in real time and shift its center of gravity in anticipation.

The number of degrees of freedom (DoF) defines the number of different ways in which a part of the robot's body can moved. ASIMO's head has two DoF: one for turning it up and down and one for rotation.

In total, each arm has five DoF. The shoulder has three DoF for bending up and down, moving side to side, and for rotation. The elbow and the wrist have one DoF respectively; ASIMO's hand has one DoF for grasping movement.

Each of ASIMO's legs has six DoF: three DoF in the hip joint, one in the knee joint, and two in the ankle joint. This high versatility makes ASIMO flexible for operating in a human environment.

## Specifications

- Weight: 43 kg
- Walking speed: 0–1.6 km/h
- Walking cycle: Cycle adjustable, stride adjustable
- Grasping force: 0.5 kg/hand (five-finger hand)
- Actuator: Servomotor, harmonic speed reducer, drive unit
- Control unit: Walk/operating control unit, wireless transmission unit
- Sensors:
  - Foot: six-axis foot area sensor
  - Torso: gyroscope and acceleration sensor
  - Power selection: 38.4 v/10 ah (Ni-MH)
  - Operating selection: Workstation and portable controller

For more information, contact:

Honda Motor Co., Ltd. ASIMO business room
TEL: 03-5412-1235, FAX: 03-5412-1754
ASIMO Internet Executive Office
Email: asimo@honda-asimo.com

# Summary

Creating an autonomous robot capable of accomplishing prolonged, complex, and dynamically changing tasks in the real world is a key challenge for the next generation of autonomous robots. To perform complex activities effectively, robots must possess rich perceptual capabilities to recognize objects and places and interactively communicate with people. Robots must also reason about and manage concurrent tasks and subtasks while executing them. Moreover, for prolonged success, robots must improve their control routines based on experiences to adapt to environment and task changes. Computational principles have been developed that enable autonomous robot control systems to accomplish complex and diverse tasks in the real world. Future work focuses on developing advanced perceptual, learning, and adaptation capabilities, as well as planning mechanisms for robotic agents.

## Experience-Based Robot Learning

Implementing and fine-tuning the control software of autonomous robots is a laborious and error-prone task. Further research will develop, extend, and apply various forms of experience-based learning to facilitate the software development process.

## Robot Action Planning

Robot action planning is the computational process of generating and revising high-level robot control programs based on foresight. Research has a goal to equip autonomous robot controllers with robot action planning capabilities that enable them to perform better than they possibly could without these capabilities. Robot action planning concentrates on three aspects:

- Methods for robot action planning (such as probabilistic, prediction-based schedule debugging)
- Realistic models for symbolically predicting concurrent reactive robot behavior
- Runtime plan adaptation for autonomous robots

## Probabilistic, Vision-Based State Estimation

Autonomous robots must have information about themselves and their environments that is sufficient and accurate enough to complete their tasks competently. Contrary to these needs, the information that robots receive through their sensors is inherently uncertain: Typically the robots' sensors can only access parts of their environments, and their sensor measurements are inaccurate and noisy. In addition, control over their actuators is also inaccurate and unreliable. Finally, the dynamics of many environments cannot be accurately modeled and environments sometimes change nondeterministically.

To cope with these problems, probabilistic state estimation modules were developed that maintain the probability densities for the states of objects over time. The probability density of an object's state, conditioned on the sensor measurements received so far, contains all the information that is available about an object. Based on these densities, robots are not only able to determine the most likely state of the objects, but can also derive even more meaningful statistics, such as the variance of the current estimate.

## Integration of Perception, Reasoning, and Action

Specifying the robot's modalities transparently and explicitly as part of a robot's plans rather than hiding them in separate modules, makes their use more effective, efficient, and robust. This enables robot control systems to generate, reason about, and revise different modalities. The controllers can also synchronize the robots' different capabilities and use control structures to make their use flexible and robust. The idea of having a single language for controlling and synchronizing the physical actions, image processing, planning, and communication capabilities of autonomous robots is a powerful one. It allows the application of the same planning, execution, and learning techniques to all different modalities of robots and their combination.

## Structured Reactive Controllers

Computational mechanisms were investigated that enable autonomous robots to exhibit competent, goal-directed behavior in mixed man–machine environments. Structured reactive controllers (SRCs) were developed to couple high-level reasoning with continuous low-level control processes. SRCs aim at enabling autonomous robots to accomplish nonrepetitive sets of complex jobs, which are mostly—but not always—routine, reliably and without wasting resources. SRCs are collections of concurrent control routines that specify routine activities and can adapt themselves to nonstandard situations by means of planning. SRCs execute three kinds of control processes: routine activities that handle standard tasks in standard situations, monitoring processes that detect nonstandard situations, and planning processes that adapt, if necessary, routine activities to nonstandard situations.

## Interactive, Planning-Based Software Agents

Planning-based software agents use the transformational planning of reactive behavior and reactive execution of concurrent processes to assist supply chain management planning and control.

## Autonomous Robot Games

In the AGILO project, computational techniques were investigated to enable autonomous mobile robots to competently play robot soccer. With a team of four Pioneer I robots, all equipped with CCD camera and a single board computer, robots have participated in all international middle-size league tournaments from 1998 until 2001. The key software methods employed by the Rob Cuppers software are:

- Vision-based cooperative state estimation for dynamic environments
- Synergetic coupling of programming and experience-based learning for movement control and situated action selection
- Plan-based control of robot teams

# 9

# Conclusions

WITH A MIXTURE OF ANTICIPATION AND DREAD, SCIENCE fiction has portrayed machines capable of thinking and acting for themselves but what was once the realm of fiction has now become the subject of serious debate for researchers and writers. Stanley Kubrick's groundbreaking science fiction film *2001* shows HAL, the computer aboard a mission to Jupiter deciding (itself) to do away with its human copilots. Sci-fi blockbusters such as *The Terminator* and *The Matrix* have continued the catastrophic theme portraying the dawn of artificial intelligence (AI) as a disaster for humankind.

Science fiction writer Isaac Asimov (Figure 9.1) anticipated a potential menace. He speculated that humans would have to give intelligent machines fundamental rules to protect themselves.

- A robot may not injure a human being or, through inaction, allow a human being to come to harm
- A robot must obey orders given it by human beings except where such orders would conflict with the First Law
- A robot must protect its own existence as long as such protection does not conflict with the First or Second Law.

Later Asimov added a further rule to combat a more sinister prospect: "A robot may not injure humanity, or, through inaction, allow humanity to come to harm."

Will machines ever develop intelligence on a level that could challenge humans? While this remains a question for argument, one thing is certain: computing power is set to increase dramatically in coming decades. Moore's Law, which states that processing power will double every 18 months, is set to continue for at least the next 9 years, and

■ **Figure 9.1**   *Writer Isaac Asimov and Joe Engleberger, "Father of Robotics," share views.*

quantum computers, though poorly understood at present, promise to add new tools to AI that may bypass some of the restrictions in conventional computing.

## Predicting the Future

AI researchers have long since abandoned hope of applying simplistic laws to protect humans from robots. For real intelligence to develop, machines must have a degree of independence and be able to weigh up contradictions for themselves, breaking one rule to preserve another, which would not fit with Asimov's laws. Conventional evolutionary pressures would determine whether machines become a threat to humans. They will only become dangerous if they are competing for survival, in terms of resources for example, and can match the humans' intellectual evolutionary prowess.

Surprisingly, some experts would welcome the possibility of machines' taking over from humans. The majority of significant human evolution has taken place on a cultural level and, therefore, replacing biological humans with mechanical machines capable of far greater learning and cultural development is the next logical step in evolution. Of course, there are more immediate threats to think about and combat, such as global warming, ocean pollution, war, and world overpopulation. However, the possibilities of artificial intelligence should not be completely ignored.

Humans today do not appear competent to solve many problems that we face. One solution is to make ourselves smarter—perhaps by changing into machines. Dangers exist in doing this—but these must be weighed against the dangers of not doing anything at all.

## AI Is Everywhere

In the 1940s, Thomas Watson, the head of IBM, famously predicted the world demand for computers might be as high as five. AI has had its share of off-target predictions as well. AI researchers in the 1950s predicted that a computer would be the world chess champion by 1968. It took a few more decades than that to achieve this goal.

But AI experts predict that by the middle of the century, intelligent machines will be all around us. In fact, they point out that AI already pervades our lives. Fuel injection systems in our cars use learning algorithms, and jet turbines are designed using genetic algorithms—both examples of AI, says Dr. Rodney Brooks, the director of MIT's artificial intelligence laboratory.

"Every cell phone call and e-mail is routed using artificial intelligence," says Ray Kurzweil, an AI entrepreneur and the author of two books on the subject, *The Age of Intelligent Machines* and *The Age of Spiritual Machines*.

"We have hundreds of examples of what I call narrow AI, which is behavior that used to require an intelligent adult but that can now be done by a computer," Kurzweil says. "It is narrow because it is within a specific domain, but the actual narrowness is gradually getting a bit broader," he adds.

Today, AI is about at the same place the personal computer industry was in 1978. In 1978, the Apple II was a year old and Atari had just introduced the 400 and 800. The choice of personal computers was pretty limited and what they could do was also relatively limited by today's standards. However, this metaphor may undersell AI's successes: AI already is used in pretty advanced applications including helping with flight scheduling or reading X-rays.

But the popular conception of AI (as seen with HAL in *2001*, Commander Data in *Star Trek*, and David in the film *AI*), is not far away. Kurzweil believes that within 30 years, we will have an understanding of how the human brain works that will give us "templates of intelligence" for developing strong AI. And Brooks says that by 2050, our world will be populated with all kinds of MentorBots.

Sounds outlandish? Who would have thought that by 2001, you would have four computers in your kitchen; the computer chips in coffee makers, refrigerators, stoves, and radios.

## Gradual Change

But in 2050, will our hyperintelligent coffee makers suddenly decide to kill us as HAL did in *2001*? Or will humans be made redundant by a legion of intelligent machines?

Brooks and Kurzweil believe that we will not wake up one day to find our lives controlled by all manner of AI devices. Referring to Steven Spielberg's movie *AI*, in which a company creates a robot that bonds emotionally like a child, Brooks says: "A scientist doesn't wake up one day and decide to make a robot with emotions."

Despite the rapid advance of technology, the advent of strong AI will be a gradual process. "The road from here to there is through thousands of these benign steps," Kurzweil says. It is extremely unlikely that robots will become dangerous or try to take over the world. Why won't robots become dangerous? Our MentorBots will have built in (software) protection (Asimov's first law) to prevent them from hurting people intentionally. People and MentorBots need a reason to hurt people intentionally. MentorBots don't need sex or money or food or possessions or clothes. We create them to serve us. If we didn't give them anything to do, they would have absolutely nothing to do. That quickly becomes very boring, so I think they will be happy to serve us. I believe that they will be appreciated and even loved by their owners. They will become more indispensable than anything else we own.

## Can We Build Intelligent Machines?

Instilling human intelligence or common sense into robots is one of the greatest challenges facing researchers. One possible approach is to create artificial nerve cells in software and hardware and wire them together in the same way that they are connected in the human brain.

Being able to build machines that are like animals or people is a dream that can be traced back at least as far as the 18th century. But only with the advent of the computer did the dream of intelligent machines take on a tangible form. Now, says Professor Marvin Minsky, AI pioneer and founder of the AI Lab at the Massachusetts Institute of Technology (MIT), it's only a matter of time before there are robots that measure up to people.

Other scientists offer similar predictions. Professor Hans Moravec of Carnegie Mellon University in Pittsburgh, Pennsylvania, believes that by 2010 robots will be able to move with the intelligence of small lizards. By 2020, says Moravec, machines will be as adaptive as mice; by 2030 as smart as apes; and by 2040 they will rival the full cognitive ability of human beings, having the power of imagination, and the ability to learn and modify their behavior. Eventually, says Moravec, they will be so perfect that people will implant their minds in them. Thus, by the end of the twenty-first century, human and artificial intelligence will merge, creating a new life form.

Many bold visions of this kind are based on the assumption that intelligence can be created through sheer computing power. "We are looking for the gold of the Incas, but we haven't even discovered America yet," cautions Professor Christoph von der Malsburg, who designs software that recognizes faces at Bochum University in Germany and the University of Southern California at Los Angeles. On the one hand, he says, modern computing is trying to make models of the brain, despite the fact that researchers do not yet understand how it works. On the other hand, AI researchers are slowly realizing how difficult it is to generate reliable behavior in a natural environment. To put it differently: In the laboratory, it may be possible to build a machine that recognizes faces, navigates a path, or takes hold of objects. But the real world is incomparably more complicated—and yet people manage to find their way around in it.

## The Evolution of Artificial Intelligence

Whereas the first generation of neural networks in the 1980s used very simple artificial nerve cells, temporal dynamics played an important role in the second generation (1990s). Neurons were no longer static, but instead operated with pulsed signals like their natural counterparts. Dependence allowed them to process input patterns far more complex than those that could be handled by static neurons. The third generation, which has evolved in the last few years, is referred to as *neurocognitive* because it takes into account knowledge concerning the organization of brain functions. Thus, in addition to the input of a certain visual pattern, the neurons in systems developed by Siemens researchers also receive data from other parts of the brain, such as the infer-temporal cortex.

Researchers are only slowly feeling their way around the question of what intelligence is. "There is no comprehensive theory of intelligence," says Professor Helge Ritter, a neurocomputing specialist at

Bielefeld University in Germany, who, together with his colleagues, is currently teaching a robot to recognize language and gestures. One thing, at least, is clear: Human intelligence can be traced back to the large number of specialized functions in our brains. We can identify objects of all kinds. We can move around without bumping into things. We can recognize the feelings of others and express our own emotions. We learn from experience. We plan our future. All of this is based on the complicated interactions that take place between numerous parts of our brain. But because researchers are still a long way from understanding how these parts of the brain act in concert, and because each part is itself extremely complex, the builders of MentorBots still have to limit themselves to small units of intelligence.

Despite all the advances they have made in recent years, the two teams of Siemens researchers admit that imbuing robots with intelligence is an arduous task. The main snag is that the environment in which we live is more diverse than one first realizes, says Dr. Gisbert Lawitzky, who has been involved in developing the cleaning robot. Every chair leg that we can easily avoid is an obstacle for a machine. "To a certain extent, you start to feel humbled by human intelligence," says Lawitzky. Even though they do not see any theoretical impediments to developing a truly autonomous MentorBot, the researchers are far from piecing together the components of intelligence into a machine that senses, plans, and acts perfectly, communicates with people, and carries around a sort of picture of the surrounding world in its head.

## Say It with Feeling

Dr. Bernd Kleinjohann, creator of the Machine with Emotionally Extended Intelligence (Mexi), is not particularly concerned whether MentorBot's emotions are genuine or pretended. What are important, he believes, are the feelings that the artificial head triggers in people. "People project their emotions onto technical devices they interact with," he observes. Although robots or artificial characters on a screen—so-called *avatars*—do not have real feelings themselves, they trigger emotional reactions in people. And this can be used, for example, to design better user interfaces. This new discipline is called *robotic user interfaces*, and the objective behind it is not to build robots and avatars that resemble people, but to develop synthetic creations that can bridge the gap between human needs and the information present in the computer world.

Neglected by cognitive researchers until recently, emotions now seem to be essential to the success of AI, a field that has disappointed many since its promising birth in the 1960s and 1970s. The Affective Computing research group at MIT is proceeding from the assumption that emotions are important for the ability of intelligent machines to make flexible and rational decisions.

At Siemens Corporate Technology (CT), Dr. Stefen Schoen's colleague Heinz Bergmeier is currently developing avatars for a future UMTS chat application. The new figures resembles amusing cartoon characters and can depict the feelings of chat participants. They work in much the same way as the emotions of many cell phone users.

Using a slide control, the user can select from up to thirteen different emotional states to transfer to his or her cell phone partner's phone. If the participants like each other, they can even go into a private room and have the penguin and the tortoise kiss by pressing a button. Obviously, it is easy to call forth emotional reactions in people. A yapping plastic dog like Sony's toy robot Aibo or a cartoon character on a cell phone display are all it takes. But could the reverse be true? Would it be possible for machines to recognize and use emotions? To date, achievements in this area have been modest. Researchers at the University of Munich have used a computer to interpret 80 percent of human gestures, but the number of gestures was small and performed by actors.

## Autonomous Robots

Autonomous robots are more common in social and home environments, such as a pet robot in the living room, a service robot at office, or a robot serving people at a party. The robot can identify people in the room, pay attention to their voices and look at them to identify visually, and associate voice and visual images, so that highly robust event identification can be accomplished. These are minimum requirements for social interaction.

Sound is essential to enhance visual experience and human–robot interaction, but usually most research and development efforts are made mainly towards sound generation, speech synthesis, and speech recognition. The system also includes face identification, speech recognition, focus-of-attention control, and sensory motor task in tracking multiple talkers. The delay of tracking is 200 msec. Focus of attention is controlled by associating auditory and visual streams with using the sound source direction and talker position as a clue. Once an association is established, the humanoid keeps its face to the direction of the associated talker.

Accelerating technological change has stretched human adaptability to the limit. Since the dawn of humanity, our ancestors lived in tribal villages. Our instincts are tuned for such conditions. The many cultural innovations of the agricultural civilizations and their successors allowed them to dominate and absorb their tribal neighbors, at a cost. It became harder and harder, and took longer and longer, to condition new citizens for the increasingly strange new ways of life. Today, schooling can consume half a working lifetime, and some people never manage to adjust. History suggests that further progress, instead of stretching us even more, can decouple us from the change. As ever more work is automated by increasingly intelligent machinery, humans could return to an approximation of their ancient patterns.

Barring cataclysms, the development of intelligent machines may be a near-term inevitability. Like airplanes, but unlike spaceships or radio, they will be a direct imitation of something already existing biologically. Every technical step towards MentorBots has a rough evolutionary counterpart, and each is likely to benefit its creators, manufacturers, and users. Each advance will provide intellectual rewards, competitive advantages, and increased wealth and options of all kinds. Each can make the world a nicer place in which to live. At the same time, by performing better than humanly possible, the robots will displace humans from essential roles. Rather quickly, they could displace us from existence. I'm not as alarmed as many by the latter possibility, since I consider these future machines our progeny—"mind children" built in our image and likeness, ourselves in more potent form. Like the biological children of previous generations, they will embody humanity's best hope for a long-term future. It behooves us to give them every advantage, and to bow out when we can no longer contribute.

But, as with biological children, we can probably arrange for a comfortable retirement before we fade away. Some biological children can be convinced to care for elderly parents; similarly, "tame" super intelligences could be created and induced to protect and support us, for a while. It is to the "wild" intelligences, however, those beyond our constraints, that the future belongs. The available tools for peeking into that strange future—extrapolation, analogy, abstraction, and reason—are, of course, totally inadequate. Yet, even they suggest surreal happenings. Robots sweep into space in a colonizing wave, but then disappear in a wake of increasingly pure thinking stuff. A "Mind Fire" burns across the universe. Physical law loses its primacy to purposes, goals, interpretations, and God knows what else.

272

## Power and Capacity

The need for computational power and memory is a logarithmic scale on both axes: each vertical division represents a thousand-fold increase in processing power, and each horizontal division a thousand-fold increase in memory size. General-purpose computers can imitate other entities at their location in the diagram, but the more specialized thinkers cannot. A 100 million MIPS computer may be programmed not only to think like a human, but also to imitate any other similarly sized computer. But humans cannot imitate 100 million MIPS computers—our general-purpose calculation ability is below a millionth of a MIPS. Information handling capacity in computers has been growing about ten million times faster than it did in nervous systems during our evolution. The power doubled every 18 months in the 1980s, and is now doubling each year. A comparison between edge and motion detectors in the human retina with similarly functioning computer vision programs suggests that the retina does the job of 1,000 MIPS (million of instructions per second) of computing. The whole brain is 100,000 times larger than the retina, so is worth perhaps 100 million MIPS of efficient computation.

AI seeks to simulate human behavior in and through "intelligent" machines. Computers with sophisticated memory chips are performing tasks that involve more than just mathematical calculations—work that many researchers once thought only humans could do. This includes giving expert advice, understanding a natural language, speaking like a human, and recognizing complex patterns such as handwriting. In May 1997, a special-purpose computer nicknamed Deep Blue from IBM won a chess match against Gary Kasparov, called by some the greatest chess player ever. But if the task should change, as it usually does in research, the general machine can just be reprogrammed, while the specialized machine must be replaced.

## Navigation

The fundamental issues in effective and safe navigation of mobile indoor robots include the problem of self-localization and collision avoidance during the process of self-localization. A successful realization of this process has to address several sub problems. A successful system for autonomous mobile robot localization should address these sub problems (Figure 9.2).

■ **Figure 9.2**  *Navigating an office.*

One approach to position estimation is a fine-grained grid-based implementation of Markov localization. The paradigm of Markov localization is based on a general framework for state estimation and applies probabilistic representations for the robot's location, the outcome of actions, and the robot's observations. The basic principles of the technique are summarized in the following sections.

## Global Localization

Global localization means the ability to estimate the position of a robot *without knowledge of its initial location* and the ability to localize a robot if its position is lost. In contrast with most existing position estimation techniques, implementation of Markov localization can represent multimodal probability distributions. Since Markov localization estimates the location of the robot in the entire environment, it is able to detect situations in which the position of the robot is lost. Therefore, our technique meets all requirements for global localization.

## Robust Localization

Most localization techniques rely on a static model of the environment, so a robust localization in dynamic environments requires the ability to estimate the position of a robot even when a significant number of the observations cannot be matched with the map—appearance of typical indoor environments may change significantly due to furniture that is moved around, opened or closed doors, or people that are around the robot. To localize a mobile robot reliably in such dynamic environments, this approach introduces a technique that filters sensor data.

The filters are designed to eliminate the damaging effect of sensor data corrupted by unordered dynamics.

## Active Localization

The efficiency of global localization can be improved by actively differentiating among different possible locations. Actively controlling the actuators of a robot to localize it is especially rewarding whenever the environment possesses relatively few features that enable it unambiguously to determine its location. Key open issues in active localization are "where to move"' and "where to look" so as to best localize the robot. To derive the means for determining the best action with respect to localization, a decision-theoretic extension of Markov localization is introduced. By choosing actions to minimize the expected future uncertainty, this approach is capable of actively localizing a mobile robot from scratch.

## Safe Localization

Active localization raises the problem of how safely to control a robot, especially during the process of active global localization. Most existing approaches to safe navigation rely on a purely sensor-based, reactive collision avoidance. To overcome the limitations of this paradigm, a method for position estimation has been combined with a reactive collision avoidance technique. The resulting hybrid approach to collision avoidance differs from previous approaches in that it considers the dynamics of the robot and avoids collisions with invisible obstacles even if the robot is uncertain about its position.

# Moral and Social Issues

Would it not be better for society if we totally banned the design and manufacture of MentorBots? We might go further and ban all research using MentorBots. Is there not a certain amount of species pride here, that we *Homo sapiens* think we have the right to dictate which sapient life forms should exist on our planet? Clearly, if MentorBots already existed, and governments decided to wipe them all out, then this would amount to a serious crime on a par with crimes against humanity, or the eradication of nations. If robots were consulted, then surely they would want to live. Besides, we might benefit from another sapient life form on our planet.

In other words, in creating robots, are we taking on the role of God? The simple response to this is that every time a couple decides to have children, they are also taking on this godlike role in bringing forth a new person into the world. If it is OK to bring new humans into the world, why is it not OK to bring a new robot into the world? Of course, one could argue that the book of Genesis tells us that God created the world for humans to live in, and not necessarily for MentorBots. The simple reply to this is: How do we know that God does not desire the creation of robots just as much as people?

## Are Regulations Needed for MentorBots?

Initially, the manufacture of robots might be expensive, but this expense would soon decrease with advancing technology. Furthermore, maintenance costs will in time be lower than those of humans. Do we want the planet to be totally overcome by robots, with many more robots than humans? Probably not, initially. The crucial question is whether families will bring up robots. Would a robot, when it was young, have an owner? Or is the whole notion of ownership relevant for MentorBots? We would not really say that we own our children.

## How Will MentorBots Affect Values?

Over 10 percent of the population is unemployed. How will the addition of MentorBots affect this problem? Of course, it is possible that society will adapt so that there are many more employment positions that would be specific to humans and difficult for robots to do. Meaning and purpose have, in human history, often been related to the quest for survival, from the bare necessities of food, clothing, and shelter, to the more sophisticated levels of today's society of the car, the television, and the mortgage. But if we have a society in which not only do we already have most of our needs met, but where robots can meet new needs much faster than humans can, meaning and purpose will have to come from elsewhere than our employment.

But there are less ephemeral sources of meaning and purpose. Obviously, prayer and adoration is one avenue. If there is nothing for us to do, then we can always worship God. Indeed this is the picture we have of our future lives in heaven—but we might get bored with a 10-hour day of praising God. There would also be plenty of scope for personal development and education. Given a lifetime of 150 years (according to the latest scientific research,) this would give us plenty of time to visit all sorts of places, to study many cultures and subjects. Of course, this would mean real education, a love of learning for its own sake, rather than to achieve wealth or fame. There would be plenty of scope to learn to be better at loving our neighbors, at caring for our families and friends: In a sense, love does make the world go round, and it is no accident that stories of self sacrifice for others move us greatly and form an important part of our culture.

## Will MentorBots Take Human Jobs?

We are already seeing machines taking up more and more aspects of employment, and this trend is likely to continue. Many human occupations are relatively simple, and can be performed by not very intelligent machines. At present this raises the possibility of humans' doing the more skilled work, with the more mundane work being done by machines. But once MentorBots become as intelligent as us, and even more intelligent, then employment patterns will begin to shift. Just as in science fiction, we will have robots taking a greater part in the workplace.

The most important reason for building MentorBots is to help us enjoy life and to relieve us of many of the mundane tasks that we all face every day. Thus, MentorBots will be able to take over for us and perform those mundane tasks for us—but, only if they have our shape and capabilities.

# The MentorBot Building Business

The MentorBot building business will become one of the greatest new industries of the twenty-first century. By the middle of the century and perhaps sooner, it will rival the automobile industry in size and importance. Indeed, I would compare the MentorBot business today to the automobile business of 1900. Many universities and corporations are even now trying to build successful MentorBots. Very soon MentorBots will be available for sale—and in a price range where many people will be able to afford them. While many people will fear for their jobs when MentorBots appear, I believe that the MentorBot industry will create more jobs than it destroys—just as the computer industry did.

## Uniqueness of MentorBots

MentorBots will be a unique product because they will be the first product that can build itself! This ability should lead to a lower-cost product because of low labor costs. Not only will they build themselves, but they will also be able to tell you what is wrong when something inside them fails. In some cases, MentorBots will be able to repair themselves, and, at the very least, repair other MentorBots—much as doctors repair humans.

## Super-intelligence

Many people have probably already realized that super-intelligence already exists—it is called the Internet. Clearly, all that is needed is to interface MentorBots to the Internet and they will rapidly become super intelligent too. The Loebner prize is given to a computer program that can pass the Turing test, which if passed is considered an indication of intelligence. To date, computers have not done very well. However, soon computers (and more specifically MentorBots) will become super-intelligent due to their interface to the Internet. They will be able to converse in many if not all languages and will have access to unlimited information via the Internet.

## What Can a MentorBot Do?

The short answer is: just about anything a person can do. It will have no need to eat or breathe or to perform other bodily functions and will therefore not be equipped to do so. However, in the future, there will doubtless be versions that will be able to have sex. Although not useful for the MentorBot, it will clearly be of interest to some buyers.

One very important function a MentorBot will perform for you is to be a 24-hour security guard for your home or business. Since it need not sleep, it will always be awake and can sound the alarm in case of burglary or fire. It could also phone the police or fire department and tell them where the emergency was (something your trusty dog cannot do). It will also be able to converse intelligently with you about any subject and it will of course remember all of its previous conversations with you. It will become your best friend. It will probably help kids with homework. MentorBots will provide in-home care for the elderly or infirm, thus reducing the cost of medical care.

An interactive robot named CosmoBot is teaching children with speech, language, and other developmental disabilities how to express themselves. Built to withstand active play, CosmoBot looks like a spunky sidekick from a science fiction movie, with fully mobile appendages, motorized wheels beneath its feet, and a mouth that moves. The robot captures attention by mimicking a child's movements and voice, and can guide the child through educational and therapeutic activities under the direction of a therapist.

The robot was developed by Corinna Lathan of AnthroTronix, Inc., in College Park, Maryland, with support from the National Science Foundation's (NSF), Small Business Innovation Research (SBIR) Program, and the Department of Education's Rehabilitation Engineering Research Center (RERC) on Telerehabilitation.

A user controls CosmoBot with wearable sensors, by voice, or with a mission control station, depending on the therapy. The toy also accepts updated software and inputs from the Internet.

"The interactive technology behind CosmoBot has roots in military research," said NSF program manager Sara Nerlove. "To move this technology to the arena of language development for children with severe or multiple disabilities is a bold and innovative step."

Early trials with CosmoBot, and its predecessor JesterBot, suggest that the toys motivate children who have developmental disabilities and prompt spontaneous and imaginative responses in speech and language therapy. The company has plans for further testing and development.

## Artificial Life

The relatively new field of *artificial life* attempts to study and understand biological life by synthesizing artificial life forms. To paraphrase Chris Langton, the founder of the field, the goal of artificial life is to "model life as it could be so as to understand life as we know it." Artificial life is a very broad discipline that spans such diverse topics as artificial evolution, artificial ecosystems, artificial morph genesis, molecular evolution, and more. This offers a nice overview of the different research questions studied by the discipline. Artificial life shares with AI its interest in synthesizing adaptive autonomous agents. Autonomous agents are computational systems that inhabit some complex, dynamic environment, sense and act autonomously in this environment, and thereby realize a set of goals or tasks that they are designed for.

The goal of building an autonomous agent is as old as the field of AI itself. The artificial life community has initiated a radically different approach towards this goal, which focuses on fast, reactive behavior, rather than knowledge and reasoning, as well as adaptation and learning. Biology, and more specifically the field of ethology, attempts to understand the mechanisms that animals use to demonstrate adaptive and successful behavior, largely inspire its approach.

Autonomous agents can take many different forms depending on the nature of the environment that they inhabit. If the environment is the real physical environment, then the agent takes the form of an autonomous robot. Alternatively, one can build 2-D or 3-D animated agents that inhabit simulated physical environments. Finally, so-called "knowbots," software agents, or interface agents are disembodied entities that inhabit the digital world of computers and computer networks. There are obvious applications for all these types of agents. For example, autonomous robots have been built for surveillance, exploration, and other tasks in environments that are inaccessible or dangerous for human beings. There is a long tradition of building simulated agents for training purposes, and more recently, interface agents have been proposed as a mechanism to help computer users deal with work and information overload.

The entertainment industry is a potential application area of agent research that has received surprisingly little interest so far. This area may become much more important in the coming years, since the traditional main funding source of agent research, the defense industry, has been scaling down. Entertainment is an extremely large industry that is only expected to grow in the near future. Many forms of entertainment employ characters that act in some environment. This is the case for video games, simulation rides, movies, animation, animatronics, theater, puppetry, certain toys, and even party lines. Each of these entertainment forms could potentially benefit from the casting of autonomous semi-intelligent agents as entertaining characters. Entertainment is a fun and challenging application area that will push the limits of agent research.

## Entertainment Robots

Several forms of commercial entertainment currently incorporate automated entertaining characters. Most of these characters are extremely simple: They demonstrate very predictable behavior and do not seem very convincing. In particular, this is the case for characters with whom a person can interact in real time, for example, video game characters. When automated characters show sophisticated behavior, it is typically completely mechanical and noninteractive and the result of a painstaking and laborious process. An example of the latter is the behavior of the dinosaurs in the movie *Jurassic Park*.

A few exceptions have emerged: A number of researchers have applied agent technology to produce animation movies. Rather than scripting the exact movements of an animated character, the characters are modeled as agents, which perform actions in response to their perceived environment. Reynolds modeled flocks of birds and schools of fish by specifying the behavior of the individual animals that made up the flock. The same algorithms were used to generate some of the behavior of the bats in the movie *Batman*. Terzopoulos has modeled very realistic fish behavior including mating, feeding, learning, and predation. His models have been employed to make entertaining, short, animated movies.

Building these entertaining agents requires the same basic research questions that are central to all agent research namely perception, action selection, motor control, adaptation, and communication. The agent has to perceive its environment, which is often dynamic and unpredictable, especially when a user is able to affect it. It has to decide what to do next so as to make progress towards the tasks it is designed to achieve. Relevant actions have to be translated into concrete motor commands. Over time, the agent has to change and improve its behavior on the basis of its past experience. Finally, the agent has to be able to communicate to other agents in the world, both human and artificial ones. The key problem is to come up with an architecture that integrates all these functionalities and results in behavior that is fast, reactive, adaptive, robust, autonomous and, last but not least, "lifelike." Lifelike behavior is no mechanistic, no predictable, and spontaneous.

Finally, building entertaining agents requires the agent researcher to think more about the user. The researcher is forced to address the psychology of the user: How will the typical user perceive the virtual characters? What behavior will he engage in? What misconceptions and confusing situations may arise? Other disciplines, such as human–computer interaction, animation, sociology, literature, and theater are particularly helpful in answering these questions. For example, animation teaches us that users typically perceive faster moving characters as being younger, upbeat, and more intelligent. Literature and theater teach us that it is easier for users to quickly grasp stereotype characters.

## Companion Robots

The underlying goals of compelling interaction and maximal autonomy have remained constant throughout the creation of robots. However, each succeeding robot has been the product of a complete redesign based on lessons learned from prior robots. Although some technical aspects have remained unchanged, such as the programming language and robot chassis, virtually all else has evolved in an effort to improve the autonomy and interactivity of the robots. From this unique position of an established trajectory of real-world interactive social robots, studying the evolution of this robot series promises to uncover valuable information for the young science of social robotics (Figure 9.3).

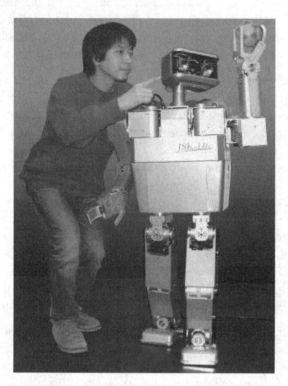

■ **Figure 9.3**  *Interact with humans.*

## Intelligent Robot Trends

An intelligent robot is a remarkably useful combination of a manipulator, sensors, and controls. The current use of these machines in outer space, medicine, hazardous materials handling, defense applications, and industry is being pursued with vigor but little funding. In factory automation, such robotics machines can improve productivity, increase product quality, and improve competitiveness. The computer and the robot have both been developed during recent times. The intelligent robot combines both technologies and requires a thorough understanding and knowledge of mechatronics.

Today's robotic machines are faster, cheaper, more repeatable, more reliable, and safer. The knowledge base of inverse kinematic and dynamic solutions and intelligent controls is increasing. More attention is being given by industry to robots, vision, and motion controls. New areas of usage are emerging for service robots, remote manipulators, and automated guided vehicles. Economically, the robotics industry now has more than a billion-dollar market in the United States and is growing.

Intelligent robots are an ideal, a vision. All one has to do to see the intelligent robot model is to look in a mirror. Ideally, all intelligent robots move dexterously, smoothly, precisely, using multiple degrees of coordinated motion, and do something like a human but that a human now doesn't have to do. They have sensors that permit them to adapt to environmental changes. They learn from the environment or from humans without making mistakes. They mimic expert human responses. They perform automatically, tirelessly, and accurately. They can diagnose their own problems and repair themselves. They can reproduce, not biologically, but by making other robots. They can be used in industry for a variety of applications. A good intelligent robot solution to an important problem can start an industry and spin off a totally new technology. For example, imagine a robot that can fill your car with gas, mow your lawn, a car that can drive you to work in heavy traffic, a machine that repairs itself when it breaks down, a physician assistant for microsurgery that reconnects 40,000 axons from a severed nerve.

The components of an intelligent robot are a manipulator, sensors, and controls. However, it is the architecture or the combination of these components, the paradigms programmed into the controller, the foresight and genius of the system designers, the practicality of the prototype builders, the professionalism and attention to quality of the manufacturing engineers and technicians, that makes the machine intelligent.

Just where is the intelligence in an intelligent robot? It is in the controller, just as the intelligence of a human is in the neural connections of the brain. However, it is only possible to see this intelligence through some action, just as it would not be possible to see intelligence in a comatose human. Where does the intelligence come from? The control program and architecture provide for real-time responses to a variety of situations. If these responses are intelligent, then the robot appears intelligent.

What are the benefits of using intelligent robots? Robots can do many tasks now. However, the tasks that cannot be easily done today are often characterized by a variable knowledge of the environment. Location, size, orientation, shape of the work piece as well as of the robot, all must be known accurately to perform a task. Obstacles in the motion path, unusual events, and breakage of tools, also create environmental uncertainty. Greater use of sensors and more intelligence should lead to a reduction of this uncertainty and, because the machines can work 24 hours a day, should also lead to higher productivity. More intelligence could also lead to faster, easier setups and reduced cycle times. More intelligence should also lead to faster diagnosis of problems and better maintenance for the systems. Finally, there is the fact that to remain internationally competitive, the best technology usage is required. Waste of human or material resources is too expensive for industry and for society.

## Mechatronics

Mechatronics is a methodology used for the optimal design of electromechanical products. The mechatronics system is multidisciplinary, embodying four fundamental disciplines: electrical, mechanical, computer science, and information technology. The mechatronics design methodology is based on a concurrent, instead of sequential, approach to design, and the use of the latest computer tools, resulting in products designed right the first time. Mechatronics covers modeling and simulation of physical systems; sensors and transducers; actuating devices; hardware components; signals, systems, and controls; real-time interfacing; advanced applications; and case studies.

## New Robot Manipulator Designs

In 1985, the four common types of robot manipulators were the Cartesian, cylindrical, spherical, and vertically articulated or anthropomorphic designs. Then the horizontally articulated or selective compliant articulated robot for assembly (SCARA) was introduced. In 1995, a totally different design, a tricept, Stewart platform was displayed by Comeau Technique Ltd. at the Robot and Vision Exhibition. It was advertised as being as flexible as a robot, as precise as a machine tool, and strong as a press. It seemed ideal for press fitting bearings and for other tasks requiring thousands of pounds rather than tens or hundreds of pounds of force.

The rotational speeds of robot manipulator links of 240 degrees/second are typical. For a 1-meter joint length, this would produce linear speeds of 4 meters/second. The overall cycle time is usually more important than individual link speeds. In a great variety of applications, robots are easily made as fast or faster than humans.

Many manufacturing applications have emerged that can't be successfully performed without robots. In the electronics industry, miniaturization is driving the demand for robots. "The trend toward further miniaturization of products like pagers, cellular phones, and two-way radios makes it virtually impossible for humans to repeatedly place, weld, or solder components accurately," according to Jim Hager, Site Manager, Motorola Manufacturing Systems, Boynton Beach, FL. "Good robotic systems can handle these tasks and help Motorola achieve Six Sigma quality."

Both industrial robots and automated guided vehicles are potentially dangerous since they move. Industrial robots in the United States have killed people. Safety requires administrative controls, engineering controls, and training. Administrative controls such as the use of the equipment restricted to qualified personnel, proper maintenance, and management insistence on safe operation, is vitally important. Engineering controls such as protective fences with safety interlocks on entrances, pressure-sensitive mats and light curtains, all properly installed and maintained, are also more commonly used than a decade ago. Training is also important and should not be overlooked, especially when a company is downsizing. Safety is not something that can be relaxed. However, more safety features and self-diagnostics can be built into robots and work cells. Also, the use of simulations to discover interferences and potential collisions is a step in the direction of safe application.

An industrial robot is easy, but putting it into an intelligent work cell requires much more than the robot. Important accessories such as grippers, process tooling, safety devices, programmable logic controllers, and simulation programs are needed to make robots easier to use.

The control system is the set of logic and power functions that allows the automatic monitoring and control of the mechanical structure and permits it to communicate with the other equipment and users in the environment. Open architecture control refers to software designs that can use or be used with products from a variety of manufacturers. The move toward open architecture controls is relatively recent but follows the trend in computers.

Even though industrial robots are position control devices, the path between position points can be extremely important, for example, in seam welding. Since any moving system is described by a dynamic differential equation according to Newton's Second Law, the dynamic solution must also be determined in the design of a robot. This dynamic solution should be used in the design of the control system. It may not be obvious that a control system as simple as that of a robot manipulator cannot be theoretically proven to be stable. However, the dynamic system is nonlinear and subject to noise from various sources.

Criteria for practical stability rather than optimal stability are used today. A variety of motion control solutions have been developed but a greater understanding of nonlinear systems is needed.

For the industrial robot to be intelligent and adapt to changes in its environments such as part location, orientation, size, and shape, sensors are needed. Vision is the most powerful senser for humans, and machine vision also adds adaptability to industrial robots, which makes them intelligent. Many robot manufacturers now offer integrated vision and robotic systems.

In the design of a robot work cell, a 3-D simulation permits one to observe interference, avoid collisions, and determine the feasibility of an operation. In some modern simulation software, once a series of motions are selected, the robot code generator program can translate the motions into robot programming language automatically and download this program to the robot. This is a major simplification and improvement.

## Realism in Robotics

New technologies go through a pattern of usage starting from zero, then increasing perhaps too far, then coming down, then reversing, and going steadily upward until they reach a downward turn at the end of their useful period. When a technology is first introduced, we may expect more than it can deliver. This period has been called the *Age of Overexpectation*. Following is a period of disillusionment in which less is expected than the technology can deliver.

This period is called the *Time of Nightmare*. Finally, reality sets in and we learn to expect only what the technology can deliver—the Age of Realism. The industrial robot has now reached this stage. The United States has a solid base of nearly 92,000 successful installations. Broadening the robot definition to include automated guided vehicles, remote manipulators that must be supervised by a human as well as totally programmable robots, and a growing interest in personal and service robots has strengthened the technology base even more.

Robotics has developed into a solid discipline that incorporates background, knowledge, and creativity of mechanical, electrical, industrial, and computer engineering and other engineering and scientific fields. However, there are still many challenges, unsolved problems, and needed inventions.

## On Robot Autonomy

The first requirement of a robot operating in a public space is safety, both for the general public and for the robot itself. At the heart of the matter is the robot's method for avoiding collisions, which must be especially robust, since the robots operate without supervision. It is notable that the collision-avoidance code on these robots is by far the least changed over the course of their existence, confirming the functionality of the initial simple design. The robots use ultrasonic range-finding sensors (sonar) to detect obstacles, and move around them reactively, each cycle choosing the appropriate motion vector to take based strictly on the most recently available sensor data, along with restrictions on how far the robot is allowed to move out of its ideal (no-obstacle) trajectory. The code is extremely simple, with no explicit mapping or modeling of the world or of the sensors themselves. It is also easy to understand, and because of the lack of internal state, easy to debug.

Because of the limited accuracy of sonar at close range, robots will occasionally become stuck when they approach a wall too closely. Given the infrequency of this failure mode (less than twice per month), robots are inherently safe. But, there is a great deal more to autonomy than safety. A robot must be able to interpret its own behavior, to determine whether or not it is functioning correctly. For humans to be confident in a robot's ability to run without supervision, a robot must be able to determine on its own when a failure occurs. Early in the development of these platforms, pagers were used, which the robot could signal via electronic mail.

The ability to recognize failure and actively request help satisfies near-term requirements for autonomy. Of course, the ultimate goal is that the robot never needs to send for help at all, so self-repair becomes a second step to self-diagnosis. Initially, the robot sends for help quickly, giving up as soon as a failure is detected. Soon it begins adding diagnostic methods to reset subsystems that weren't functioning correctly. This evolves into a general method for autonomy within object-oriented architecture: Every time a task is performed or an external piece of hardware is commanded, the result is checked for validity. If invalid, the device resets and tries again.

When docking to recharge, for example, if the battery voltage fails to rise when the robot believes it is plugged in, the robot will reset the physical situation by backing out of the plug and into the hallway. Then, it will repeat the docking attempt. This "try again" policy is effective in robotics because, although the code is deterministic, sufficient nondeterminism exists in the environment that the same code may have different outcomes. This policy can be further refined with the caveat that the failure mode of an attempted task must be noncatastrophic for a retry to be possible. Robots have evolved to make increasing use of this strategy and now detect a number of abnormal situations, many of which are automatically corrected, including battery overcharging and undercharging, frame-grabber anomalies, DVD player errors, bizarre encoder values (that indicates the robot had been pushed by an external force), emergency-stop activations, and the like. Like other aspects of the robot, robot navigation and vision evolved to meet these challenges.

The same "try again" technique is also used to make the landmark searching algorithm more robust. Chips using a specific set of pink visual landmarks, one of which is three-dimensional, are used to provide corrections to simple encoder-only methods for determining location. As robots are installed in more locations, new kinds of visual landmarks are added, including edges sharp in intensity and rectangles of different color. Different methods for using them are also added, allowing multiple landmarks to be tracked simultaneously (to deal with changing lighting conditions), and using landmarks to allow the robot to localize in more directions.

Because the science of human–robot interaction (HRI) is in its infancy, it is not surprising that the robot interaction component has been redesigned over time. Even so, several qualitative conclusions have been reached. An interview with the exhibits maintenance staff of any large museum will drive home an important fact: people are basically destructive. Sometimes this is purposeful damage caused by malicious people; more frequently, curious individuals who are trying to better understand the robot cause damage.

For example, some will attempt to push the robot off course to see if it will recover. Others will push any large red emergency stop button to see what happens. Also, what attracts people varies greatly depending on the context of a particular public space. When in an "entertainment" space, such as a museum, people will be curious and attracted by new and unusual things, so the physical appearance of the robot is important. But two other characteristics produce even better results: motion and awareness. A robot in motion draws the greatest attention from nearby people.

Early indications are that some success is possible using a much more socially aggressive robot that physically approaches individuals to initiate interaction. In addition to attracting an audience, a robot must be able to retain one. Museum exhibit designers have tended to make their exhibits more interactive, often even taking on the characteristics of conversation. An exhibit may pose a question requiring the visitor to lift a panel or push a button to hear the correct answer. This is because attention tends not to stay focused through long presentations. Visitors who are involved in the exhibit stay more focused and curious about the information being conveyed.

Finally, an aid to increasing the complexity of the dialog is for the robot to have multiple ways of answering the same question so that it seems less scripted and more spontaneous, and therefore more interesting. A final lesson learned with respect to HRI involves the psychological effect of creating an anthropomorphic robot. There are strong social rules governing appropriate behavior between humans (though these rules vary between cultures and segments of society), and there are other behavior patterns that people follow when interacting with machines and computers.

A robot that looks somewhat human and has a rudimentary personality falls somewhere between human and machine. The majority of people treat a robot as part human, part machine, clearly following some modified form of human interaction. Often they will treat the robot like a human by default, getting out of its way and verbally responding to it. If they become interested in some feature of the robot, or want to investigate how it works, however, they will start treating it like a machine, ignoring its requests to move, and standing rudely in its way to see its reaction.

The main reason for a robot to display emotions is that humans expect and respond to them in somewhat predictable ways. People have a strong anthropomorphic urge and tend to attribute internal state to anything that behaves appropriately. People are also strongly conditioned to react to the emotions displayed by another person. These are powerful tendencies that robots should exploit. These reactions are entirely behavioral. People cannot discern the true internal state of another human or robot. Their responses are thus entirely dependent upon perceived behavior.

Yet if the robot is busy giving a tour it should politely ask the person to join the tour or, failing that, to please get out of the way so that the tour group can move along. Even more important than having reactions for all possible interaction contexts, it is critical that the robot's reactions are correct. If the robot begins talking to a wall or to thin air, it looks truly stupid. Just as moving safely through a crowd without hurting anyone is a basic required competence for a mobile robot, so total avoidance of stupid social interactions is a basic competence.

When a robot deliberately faces a person and says "Hello," the person is almost always both surprised and enthralled. In contrast to entertainment venues, more utilitarian spaces such as shopping centers and office buildings elicit far less pronounced reactions. In these spaces, people tend to have an agenda; they rush about and are less willing to be sidetracked by a new and entertaining creature. Early indications are that some success is possible using a much more socially aggressive robot, which physically approaches individuals to initiate interaction.

A robot also must have an adequate depth of dialog so that a human cannot immediately exhaust the robot's "conversation space," rendering the robot predictable, and therefore uninteresting. But in designing this personality, one must be as conservative as when designing obstacle-avoidance code. Making obvious mistakes, such as talking to a potted plant, will cause the robot to be completely dismissed by the audience. In the domain of autonomy, an approach to design and implementation that implicitly promotes fault tolerance is important for the long-term survival of a robot.

The basic "try again" approach works extremely well since the same code executed twice on a robotic platform will often yield different results. This approach, coupled with the ability of the robot to send pages when it needs help, makes human supervision refreshingly unnecessary. Even so, there are some types of failures that a robot cannot recover from completely, even if detection of the failure is possible. Drained batteries, burned-out fuses and light bulbs, and cooked sonar transducers have brought robots down at various points in time, and the robots simply cannot fix such failures. Mobile robots still depend on humans for their continuing existence.

Generally, people will notice if the robot fails to react to some indirect stimuli, but they will notice if the robot reacts inappropriately. In summary, the interactivity of robots has evolved along four axes: engagement, retention, dialog, and anthropomorphic/affective qualities. Although this field of research is extremely young, it is already clear that there remains great pliability in the human–robot interaction model: human biases and bigotry regarding robots are not yet strong and fixed.

The internal rate of return may be interpreted as the return on the investment in automation. The reciprocal of the internal rate of return is the payback period or the time in which the investment will be recovered. After the payback period, the automation equipment is producing wealth. Interestingly, in almost all industrial applications with sufficient production, a robot installation is nearly always feasible. The question is whether the rate of return is sufficient. An internal rate of return of 50 percent or a payback period of two years is often suggested for industrial applications. This is not the case for all robotic applications. For example, robots in most space, undersea, environmental, defense, and service applications are not yet proven technology and cannot be easily cost justified using the internal rate of return concept. These are not poor investments, but rather, like education and research, vitally important activities in the long term.

The growth of the robotics field in the United States is indicated in several ways. In 1982, the RIA indicated that 6300 industrial robots were in use in the United States with 2453 used for welding, 1060 for machine loading and unloading, 875 in casting, 1300 in material handling, 490 in painting and finishing, and the remaining 122 for assembly and other areas. According to the RIA more than 92,000 robots were at work in U.S. factories in 1999.

The number of robots in use tells one important part of the story; however, another important aspect for the United States is the technology base of trained engineers and technicians who are familiar with industrial robots. A search of the World Wide Web resulted in more than 3500 robotics references. The number of people who have interests and training in the cross disciplines of mechanical, electrical, computer, and industrial engineering mechatronics is increasing. These manufacturing engineers and technicians no longer look at a task or a machine and see only the operating machine but also appreciate that the concurrence of all components and the consensus of all the humans involved to produce a successful product. Whether this concept is called total quality, consensus management, or customer awareness, it has been an important lesson to learn.

## When Will MentorBots be Available?

Well over a 100 different groups or people are now working to develop MentorBots. Some believe that 6 percent of U.S. homes will have a

MentorBot within 12 years, and 50 percent will within 20 years. That represents a potential market of 6 million units in the next 12 years, just in the United States. It is probably the same for Japan or Western Europe, thus giving a potential of 20 million units worldwide in the next 12 years.

## MentorBot Sales 2001–2015

The basic plan is to achieve 6 percent placement of MentorBots in the developed nations of the Americas, Asia, and Europe within 12 years. We have not assumed, as may be possible, that the Japanese and European markets may actually exceed the U.S. market (although with different but similar percentages) in the same timeframe. In the United States, we have about 100 million households. Thus 6 percent penetration would amount to 6 million units sold by 2014. Due to the much greater popularity of robots in Japan, we believe that Japan will achieve sales at least as high as the United States, in spite of a population only about half that of the United States. Europe should be able to support sales similar to the United States. The projected sales would be as follows:

| Year | Worldwide Robot Sales | MentorBot World Sales |
| --- | --- | --- |
| 2001 | 30,000 | 900 |
| 2002 | 50,000 | 2000 |
| 2003 | 90,000 | 4000 |
| 2004 | 190,000 | 7,000 |
| 2005 | 370,000 | 12,000 |
| 2006 | 680,000 | 20,000 |
| 2007 | 1,150,000 | 35,000 |
| 2008 | 1,800,000 | 60,000 |
| 2009 | 2,750,000 | 100,000 |
| 2010 | 3,850,000 | 170,000 |
| 2011 | 5,000,000 | 300,000 |
| 2012 | 6,400,000 | 450,000 |
| 2013 | 8,000,000 | 700,000 |
| 2014 | 10,000,000 | 1,200,000 |
| 2015 | 12,000,000 | 1,800,000 |

According to the newly released survey "World Robotics 2002," produced by the United Nations Economic Commission for Europe (UNECE), "2001 was a record year for robot investment in Europe." At the end of 2001, there were about 360,000 industrial robots in Japan; 220,000 in Europe; and 100,000 in North America. The

European market has continued to grow, despite the 9/11 crisis, while sales have plummeted in Japan and the United States. For the period 2002–2005, "the world market is forecasted to grow by an annual average of 7.5 percent," says the survey, which indicates that Europe will experience a dramatic increase of its robotics population: By 2005, there should be 352,000 units in Japan; 321,000 in the European Union; and 131,000 in North America.

The survey also explains that "Household robots are now on the verge of taking off." with the emergence of robotic vacuum cleaners, automatic lawnmowers, or entertainment robots. At the end of 2001, it was estimated "that over 20,000 domestic robots, all types included, have been installed" while entertainment robots count for 150,000 units. The survey forecasts "a huge increase in sales" for the period 2002–2005, both for domestic robots (vacuum cleaning, lawn mowing, window cleaning and other types), that could reach 700,000 units, and for toy and entertainment robots that could exceed one million units (Figure 9.4).

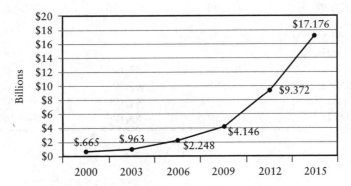

■ **Figure 9.4**  *Mobile robot sales 2001–2015.*

Although the U.S. market is healthy and growing, the United States is still significantly behind Japan in the use of industrial robots, automated guided vehicles, and mechatronics. The Mazak Corporation is a good example of a worldwide leader in advanced manufacturing technology. The company president, Teruyuki Yamazaki has stated "The greatest secret in promoting marketing activities efficiently is to get a grasp of exactly what the market requires and to develop and offer to the market those products which will sell even when there is a recession." Robots can play a significant role in improving productivity, quality, and flexibility and time to market. Table 9.1 summarizes the use of robots worldwide.

**Table 9.1**  *Robot and MentorBot Sales 2001–2015*

| Business Robots | Stock in 2000 | Sales 2001–2005 | Sales 2006–2010 |
|---|---|---|---|
| **Worldwide MentorBot Market:** | | | |
| Marketing | 20 | 100 | 500 |
| Gofer | 30 | 150 | 750 |
| Other | 50 | 250 | 1250 |
| Caregiver Robots | | | |
| Nursing | 100 | 1000 | 5000 |
| Nanny | 100 | 1000 | 5000 |
| Companion | 400 | 4000 | 20,000 |
| Domestic Robots | | | |
| Vacuum cleaning | 2000 | 10,000 | 30,000 |
| Lawn Mowing | 3000 | 15,000 | 45,000 |
| Other | 7000 | 30,000 | 90,000 |
| Misc. Use Robots | | | |
| Educational | 1000 | 10,000 | 60,000 |
| Leisure/hobby | 2000 | 20,000 | 120,000 |
| Entertainment | 5000 | 50,000 | 300,000 |
| Other | 10,000 | 100,000 | 1,000,000 |
| TOTALS | 30,700 | 241,500 | 1,677,500 |
| | | | |
| **World Market Segments:** | | | |
| MentorBots | | | |
| Japan | 10,440 | 82,100 | 570,300 |
| European Union | 8290 | 65,200 | 452,900 |
| United States | 5220 | 41,100 | 285,200 |
| Other | 6750 | 53,100 | 369,100 |

## Summary

Intelligent robots for industry make sense technically, economically, and socially. Robotic devices that increase the level of flexibility of industrial automation can directly lead to improved productivity. The feasibility of a successful implementation is high. Also, such automation is a good investment. Repetitive jobs that a robot can do, such as applying sealant to rear windows of an automobile at the rate of a million per year, or stacking boxes at the rate of 360 per hour, are unfit for humans since repetitive motions by humans lead to cumulative trauma disorders.

It appears that most advances in intelligent robots have been through "bottom-up" applied research. One application at a time is being solved. Furthermore, this advance is being funded directly by industry. The limitation of this bottom-up approach is that only low-risk technology will be developed. If research is limited to current products, where will new products come from?

There is also a need for "top-down," high-risk research and development. New ideas need to be tried. Theoretical research that can be used by everyone and may never be seen from the bottom-up use of existing technology needs to be funded by the government. A new robot could not or should not be built until the inverse kinematic and dynamic solutions are known. Theoretical advances are needed in intelligent control theory that would at least enable us to say that robots are stable and controllable and safe before millions of these devices are put out in society. Sensor integration can be attempted in a research environment much more easily that in an actual application.

Robots play golf, weld, spray paint, assemble, handle materials, load and unload machines, shear sheep, mow lawns, cut trees, pick oranges, solve Rubik's cube, play checkers, play black jack, play board games, fill cars with gas, make milk shakes, package hamburgers, deliver food to patients in hospitals, and perform brain surgery. All these have been done by prototype systems at least; some have made it to market, others are on the way.

## Futuristic Predictions

As stated 15 years ago: "The ultimate goal of the use of robots always should be to help, not hurt us. This calls for thoughtful planning, intelligent policies, and foresighted decisions on the parts of leaders in field that will use robots, from factory owners to medical researchers. We are approaching a new era in our continuing industrial evolution. It can mean freedom from all mindless work, freedom to develop our uniquely human abilities to their fullest extent, and freedom to explore the frontiers that we could never physically explore ourselves. Continued sensitivity, however, to the effects of technology on all people is of paramount importance for the wisest and happiest implementation of this new technology."

## Economic Aspects—Billion Dollar Market

Recent reports regarding the use of industrial robots are encouraging as shown in the following quotations from Donald Vincent, Executive Vice President of the Robot Industries Association:

"The North American robotics industry sold more robot in the first quarter of 1999 than in any quarter since the industry began collecting statistics in 1983. A total of 4,732 robots valued at $362.8 million were sold, which is an increase of 67% in units and 24% in value over the first quarter in 1998." "RIA estimates that some 92,000 robots are now at work in American factories and that the automotive industry accounts for at least 50% of all robot orders, with electronics, food and beverage, pharmaceutical, consumer goods, aerospace and appliance accounting for the majority of the other sales."

# Index

**304**

**305**

306

## About the Author

E. Oliver Severin is the founder and President of AcroTek, Inc., a company that creates and builds technology solutions. In 1982, Mr. Severin's company merged with RB Robots, becoming the largest producer of mobile robots in the United States. Pioneers in the mobile robot industry, the firm spawned major innovations in digital voice, lifting mechanics, security, and educational robots. Mr. Severin is the creator of several mobile robot innovations and the author of a number of white papers related to robots and the robot industry.